D0837153

PRAISE FOR Strange Angel

"[Pendle] depicts Parsons' short, spectacular life as akin to one of his early rocket tests—a brilliant flash, a quick soar, and an inevitable, erratic fizzle back to Earth. Pendle weaves a fascinating yarn, reaching from the earliest sci-fi dreams of manned space-flight to the real-life trial-and-error process that would eventually make it possible." —*The Seattle Times*

"Offer[s] new glimpses not only of the history of a lab, a science and a group of extraordinary people but also of America's rapidly changing political and cultural assumptions. . . . Parsons's story is an intriguing one, full of contradictions that seem quintessentially of their time."—*The New York Times Book Review*

"These days, the image of a typical rocket scientist is pretty much button-down. After all, sending probes to Titan and beyond is serious business. But once, there was a man named John Whiteside Parsons. He was a bright fellow, perhaps even a genius . . . a certifiable eccentric. His was a decidedly interesting life—and death." —*The San Diego Union-Tribune*

"That rarest of things: A popular science book that's a page-turner too." —*Popular Science*

"As a history of space travel, *Strange Angel* is a cornerstone. This is your book if you want to start reading up on the space age." —Ray Bradbury

"Unlike contemporaries who believed science and magic were inherently contradictory, John Whiteside Parsons considered the two endeavors complementary. . . . George Pendle brings to light both segments of Parsons' work." —*San Francisco Chronicle*

"[This] book tells a spellbinding story of a man with eccentricities that went well beyond a fascination with rocketry and included a penchant for the occult. *Strange Angel* has a strong narrative drive and reads like a novel—except that a novel has to be plausible, whereas the life of Jack Parsons, poet, magician and rocket pioneer, had no such constraint."

—*American Scientist*

"Pasadena's famous Craftsman mansions disgorge their ghosts in this rambunctiously funny, deliriously weird, and incredibly true story of a space-science pioneer turned lustful witch."

—Mike Davis, author of *City of Quartz*

"*Strange Angel* works as an engaging treatment of a time when the modern world moved at the same speed as crazed mania."

—*The Onion*

"A riveting tale of rocketry, the occult, and boom-and-bust 1920s and 1930s Los Angeles. Equally cogent in interpreting the scientific and personal facets of Parsons' alluringly scandalous and confounding life, Pendle greatly enlivens the story of rocketry."

—*Booklist* (starred review)

"Pendle vividly tells the story of a mysterious and forgotten man who embodied the contradictions of his time. Marshaling a cast of characters ranging from Robert Millikan to L. Ron Hubbard, Pendle offers a fascinating glimpse into a world long past, a story that would make a compelling work of fiction if it weren't so astonishingly true."

—*Publishers Weekly*

"This amazing book is set in a more brightly coloured universe than most scientific lives are."

—*Daily Telegraph* (London)

STRANGE ANGEL

STRANGE ANGEL

....

The Otherworldly
Life of Rocket Scientist
John Whiteside Parsons

....

GEORGE PENDLE

A HARVEST BOOK • HARCOURT, INC.

ORLANDO AUSTIN NEW YORK SAN DIEGO TORONTO LONDON

Copyright © 2005 by George Pendle

All rights reserved. No part of this publication may be reproduced or
transmitted in any form or by any means, electronic or mechanical, including
photocopy, recording, or any information storage and retrieval system, without
permission in writing from the publisher.

Requests for permission to make copies of any part of the work should be
mailed to the following address: Permissions Department, Harcourt, Inc.,
6277 Sea Harbor Drive, Orlando, Florida 32887-6777.

www.HarcourtBooks.com

Excerpts from John Whiteside Parsons' writings: Reprinted by permission of the
Estate of Marjorie Elizabeth Cameron Parsons Kimmel and Thelema Media, LLC.
Excerpts from Aleister Crowley's writings: © Ordo Templi Orientis.
Used with permission.
Excerpts from L. Sprague de Camp's letters: Reprinted with permission of the
de Camp Family, Limited Partnership, c/o Spectrum Literary Agency.

The Library of Congress has cataloged the hardcover edition as follows:
Pendle, George, 1976–
Strange angel: the otherworldly life of rocket scientist John Whiteside
Parsons / George Pendle.—1st ed.
p.   cm.
Includes bibliographical references and index.
1. Parsons, Jack, 1914–1952.   2. Rocketry—United States—Biography.
3. Occultists—United States—Biography.   4. Aeronautical engineers—
United States—Biography.   I. Title.
TL781.85.P37P46 2004
621.43'56'092—dc22   2004010666
ISBN-13: 978-0-15-100997-8   ISBN-10: 0-15-100997-X
ISBN-13: 978-0-15-603179-0 (pbk.)   ISBN-10: 0-15-603179-5 (pbk.)

Text set in Sabon
Designed by Cathy Riggs

Printed in the United States of America

First Harvest edition 2006
A C E G I K J H F D B

To my mother and father

• • • • • • • • • • • • •

"No rocket goes as far astray as man."

—Robert Lowell

CONTENTS

STRANGE ANGEL

PROLOGUE

• • • • • • • • • • • •

Moderation has never yet engineered an explosion.

—ELLEN GLASGOW, *The Woman Within*

At 5:08 P.M. on June 17, 1952, an explosion rips through the warm, lush air that blankets the city of Pasadena. Those who are closest to the explosion will say that there were two almost simultaneous blasts. But by the time the sounds reach the famed mission-style dome of City Hall some two miles away from their source, they have fused into one indistinct eruption. People turn blindly to the sky, trying to find the source of the explosion. Tentatively at first, then more confidently, they dismiss it as construction noise or demolition work. Pasadena is changing rapidly these days.

Those to the south of the city center can pinpoint the noise a little better. It seems to have come from the edge of the Arroyo Seco, the wild valley that runs along the western border of Pasadena and separates the city from the encroaching sprawl of Los Angeles. Indeed, those near the Arroyo flinch and swing their heads instinctively toward the source of the sound—Orange

Grove Avenue, that faded relic of the city's glorious past better known as Millionaire's Row.

On Orange Grove the explosion causes the magnolia trees to shudder. Heads appear out of windows and people stand frozen as the blast echoes off the few remaining white-washed mansions, ringing high over the empty lots and building sites. The sound does not come from the French manor house of John S. Cravens, former president of the Edison Company. Nor does it come from the Busch Gardens that had once played host to presidents and those even more powerful. The Macris estate, home to the reclusive oil heiress, stands unmoved. No, the sound of the explosion came from 1071 South Orange Grove, the old Cruikshank estate.

But the old Cruikshank manor has long since been torn down. Once the home of California's most prominent attorney, its place is taken now by some of the first condominiums to breach this once restricted area. They sit awkwardly alone, an air of incongruity hanging heavily on their modern frames as if membership to the street still eluded them. The explosion seemed to come from beyond them, down the long, serpentine drive that remains from the manor's glory days. It is now clear that the blast came from the estate's old coach house.

Loose sheaves of paper have been blown out onto the driveway and the surrounding lawn. Thin smoke cloaks the building. All but one of the large garage doors have been knocked from their hinges and lie askew on the ground, buckled and broken. The window frames hang glassless and limp from the wall. It is as if the building has disgorged itself. Getting closer, one can see that the doors previously enclosed a large room, although it is hard to make out the room's exact dimensions since it is clogged with debris. The heavy timber frames from the ceiling have collapsed, and the floor is covered in splintered wood, broken plaster, and the unidentifiable confetti of destruction. The side walls have been stripped of their plaster, and the exposed wooden support struts loomed like a broken rib cage. The back wall has exploded outward, revealing a shattered greenhouse slumped some twenty-five feet away. There is an acrid, chemical smell in the room. The right-side wall bulges unnaturally, blocking shut the door to an adjacent room. There is a large hole, charred black, in the middle of the floor.

The room appears to have been used as a laundry. A dented boiler squats in the corner, its water pipes buckled and bent. A large cast-iron wash tub has been ripped from its fitting and lies wedged against the wall as if it had been tossed aside by an un-interested child. Two people, a young man and an older woman, are straining to move it. It is too heavy to lift, so they try to roll it away. Finally it tumbles onto its side. Beneath its white bulk they find a pool of blood and, lying in it, the singed and broken body of a man. They pull him to the wall as carefully as they can and prop him up. He looks like a life-sized rag doll. The man's shoes have been shredded by the force of the explosion. His legs lie shattered and limp in front of him, unnaturally crooked. His white shirt is scorched black and stained red, the right sleeve flapping uselessly; there is no arm to fill it. But this is not the worst. The left side of the man's face is slack and ex-pressionless, and the right side appears to have disappeared al-together. The skin has been ripped off, exposing the white of jaw bone and teeth. One eye is open, the other appears not even to be there so covered in gore is the face. And there is the sound of groaning; despite his horrific injuries, the man is conscious.

Greg Ganci, who has just propped the body up, stands back in a daze. He is a young actor, in his mid-twenties, and has been renting one of the upstairs bedrooms for the past three weeks. He looks up and sees the hole in the ceiling that had appeared in his floor only moments before. Another boarder, Martin Fo-shaug, now comes downstairs and surveys the devastation. He called the police immediately after the explosion occurred. The older woman who had helped move the body, Foshaug's mother, goes upstairs. She has something on the stove and doesn't want it to boil over.

Ganci and Foshaug look at the devastation that surrounds them. As the smoke clears, the room seems to reveal itself as something other than a laundry. Around the body lie broken bottles made of thick dark glass, vials, flasks, and test tubes. The man continues to groan. Ganci says to Foshaug, "We've got to go and tell his wife." As he turns to leave, he notices hypodermic needles spilling out of one of the overturned trash cans. He looks at Foshaug and back down at the needles. "*We* don't want to be accused of taking drugs," he says with sudden urgency. They sweep them up and dispose of them as the sound of sirens emerges in the distance.

Reams of paper continue to waft around and out of the room, brushing up against the groaning, half-dead man. Some carry abstruse chemical formulas, sketches of molecular composition, and long streams of tables and equations. Neatly clipped newspaper cuttings singed by the heat of the explosion tell tales of blasts in shipyards and bombs placed under cars, of massive loss of life and unexplained causes. The garishly colored covers of science fiction magazines float by, ripped from their staples, tattered and torn. Other pages have strange symbols on them, pentagrams, cabalistic charts, and writing in unintelligible languages. As the sirens get louder and louder, the paper seems to swathe the crushed body like bandages.

Ganci is driving quickly along Orange Grove toward 424 Arroyo Terrace, a couple of miles to the north. The door is not opened by the wife Ganci expected to see, but rather by the mother, sixty-one-year-old Ruth Parsons, who has been taking care of the house for the summer. Her son and daughter-in-law have been staying with her for the past few weeks prior to a long-planned holiday, and their suitcases line the corridor. It is the first time that Ruth Parsons has lived with her son since he was a child. Ganci speaks, trying to catch his breath, "Mrs. Parsons," he says, "Jack has been injured in an explosion in the house." There is the slightest of pauses before Ruth Parsons screams. She covers her mouth with her hand, gasping for breath. "Oh, my

God!" she screams over and over again. Ganci helps her to a chair and sits with her for a few long minutes. As her sobs slowly quiet, Ganci speaks again, "Mrs. Parsons, he's very badly hurt. There's a possibility that he may not live." Ruth Parsons screams and groans again. Ganci, not knowing quite what to do, says he will send word about her son's progress, and leaves her.

By the time he returns to the coach house, Jack Parsons, still conscious, has been rushed to Huntingdon Memorial Hospital. The ambulance crew inform the arriving newspaper reporters that Parsons had struggled to tell them something, but the wounds to his face would not allow him to speak. They had tried, but could not find the rest of his right arm. In the coach house police investigators have their notepads out and are trying to ascertain the cause of the explosion. Ganci and Foshaug describe what happened as best they can. They were the man's tenants. They had known him only for a few months. Just outside the garage door, unaffected by the blast, are several boxes full of glass bottles labelled DANGER—EXPLOSIVES. The reporters photograph the police investigator as he works his way through the rubble; he estimates that there are enough explosives still remaining in the laboratory to "blow up half the block." He puts a call through to the army's Fifty-eighth Ordnance Disposal Unit to remove them and quickly covers some of the bottles so the camera flashes won't ignite any light-sensitive chemicals.

Some twenty or so reporters and photographers are now at the scene of the blast, walking in and around the building, reading the scraps of paper that lie strewn across the floor. A local newspaperman finds a fragment of legible handwriting. It reads, "Let me know the misery totally. And spare not and be not spared. Sacrament and Crucifixion. Oh my passion and shame."

Amidst this confusion of police and press, Marjorie Cameron Parsons, Jack Parsons' wife, returns to the coach house in a car pulling a trailer. The reporters, sensing a widow in the making, swarm around her. Stunned as she is, all she can say is that her husband was a chemist, that she is an artist, and that they were due to leave for Mexico that very day for a holiday. In fact, they were going to leave as soon as her husband had finished working with his chemicals. Some of the more inquisitive reporters lift up the tarpaulin covering the trailer. It contains canvases, paints, a record player, archery equipment, and fencing foils. Marjorie Parsons gets back into her car and drives to the hospital. However, by the time she arrives, her husband, John Whiteside Parsons, known as Jack to his friends, has been declared dead. He was thirty-seven years old.

It is 6:30 P.M. At 424 Arroyo Terrace Ruth Parsons has just been informed of her son's death. Ever since Ganci had told her of the accident, she has been drinking heavily. Now she becomes hysterical, screaming "I'm going to kill myself! I can't stand this!" Mrs. Helen Rowan, a friend who is staying at the house, tries to comfort her, but Ruth Parsons is staggering wildly around the house, and Mrs. Rowan, who is chair-bound with arthritis, can do little to stop her. Mrs. Rowan phones for a doctor and is told that a prescription for the sedative Nembutal is being sent to calm Ruth down. Mrs. Nadia Kibort, another of Ruth's elderly friends, arrives with the pills. She gives Ruth two of them. Ruth swallows them with alacrity and places the bottle of pills on the piano. Talking to Ruth slowly and calmly, Mrs. Kibort coaxes her into an easy chair opposite Mrs. Rowan. Slowly the sobs ease, and Mrs. Kibort, seeing that Ruth is calmer, goes to the kitchen to prepare some food. After a minute or two Ruth Parsons stands up, grabs the bottle of sedatives, and begins to swallow its contents. Mrs. Rowan, unable to stand, can only watch aghast. She begs her to stop, but Ruth does not heed her and continues to gulp down the pills. Realizing she is powerless, Mrs. Rowan screams for help and Mrs. Kibort hurries as quickly as she can back into the room. But by that time Ruth Parsons has slumped back into her chair. The pill bottle is almost empty. A nurse is called; however, by the time the nurse arrives, Ruth Parsons is barely conscious. A doctor follows quickly, but the barbiturates have completed their work. Mrs. Ruth Virginia Parsons is declared dead at 9:05 P.M., less than four hours after her son. When an ambulance crew arrives to remove the body, they find Ruth's dog standing guard at her feet.

Back at 1071 South Orange Grove, the reporters and police are leaving. Deadlines are approaching and reports need to be written. Word comes over the radio of Ruth's death, and the photographers scurry up the road to get pictures of this new macabre development. Once there is no one left, Ganci and Foshaug force their way into the room that had been closed off by the buckled wall. The room they step into is painted bright pink. Black lace is draped over the bookshelves, and on one wall is a ten-foot-high painting of a black devil's head with huge eyes and horns. "We better paint over that face," says Ganci to Foshaug, "or else this is going to haunt us for years to come." Rather ruefully, they begin to whitewash the devil from existence.

The paint was still drying on the demon in the garage when the next day's newspapers appeared. In spite of the fact that the chairman of the United States Joint Chiefs of Staff had warned Congress that the Soviet Union "could overrun Europe today," that the army had placed its homeland antiaircraft batteries on high alert against the possibility of a Soviet air attack, and that Ingrid Bergman had given birth to twins, the front page of the *Los Angeles Times* was emblazoned with the headline: ROCKET SCIENTIST KILLED IN PASADENA EXPLOSION.

At first glance the story seemed a straightforward, if shocking, family tragedy, as well as a terrible loss for the world of science. The newspapers sadly outlined Parsons' accomplishments and his terrible end. He had been a scientist at the California Institute of Technology (Caltech) in Pasadena and while there had worked with the famed Dr. Theodore von Kármán, the presiding genius of aeronautics. He had been one of the founders of the prestigious Jet Propulsion Laboratory (JPL) situated just a few miles northwest of the city, where he had engaged in top secret governmental work during the Second World War. He was recognized "as one of the foremost authorities on rocket propulsion" and had been a member of the American Chemical Society, the Institute of Aeronautical Sciences, the Army Ordnance Association, and the exclusive Sigma Xi fraternity. In addition to all of this, he had even dallied in the world of commerce as one of the founders of the hugely successful Aerojet Engineering Corporation, an aerospace company rich with governmental research projects.

Similarly, the reason for Parsons' death appeared relatively clear-cut. Don Harding, a criminologist involved in his first major investigation since his assignment to the Pasadena police department, found residue of fulminate of mercury, a highly combustible explosive, in a trash can at the scene of the explosion. He also found bits of a coffee tin shredded into shrapnel and theorized that Parsons had been using the tin to mix the chemical in when he had accidentally dropped it. Knowing the fulminate was so volatile, Parsons had quickly stooped down in an attempt to catch it. He had been too late. The can had hit the floor, the explosive had ignited, and Parsons' searching right arm and right side of his face had borne the full brunt of the blast. The explosion had then ignited other chemicals in the room, causing the holocaust. A man of promise and genius had been lost to a terrible accident.

Over the next few days, however, Parsons' life began to appear more complicated and the story of his death less straightforward. Most scientists of any standing leave a clear trail through their work and research—Ph.D.'s, published papers, articles, conferences attended—all carefully documented and easily traceable. But Parsons left little in the way of illumination. Despite statements given by former friends and acquaintances that Parsons studied at Caltech, when pressed, neither the police nor the newspapers could find any official record of an education past high school. Some said that Parsons' most recent work before his death had been making special effects for the motion picture industry—strange work for a man named as one of the "foremost authorities" on rocketry.

Two days after the explosion the Pasadena police added to these contradictions by announcing that Parsons had been investigated by them ten years earlier, following the receipt of an anonymous letter accusing him of "perversion" and "black magic." Although "charges of strange cultism were not substantiated," they had been noted. Parsons' house had been investigated in 1944 following a minor fire, at which time the investigating officers found "numerous books and pamphlets about a mysterious 'Church of Thelema'," along with paraphernalia suggesting that "spiritual séances" were held in the house. These revelations prompted the newspapers to search out the dead scientist's former colleagues to corroborate these facts, and soon journalists were being told of Parsons' "flair for mysticism" and of his interest in the occult, which seemed to rival his fascination with rocketry. From shock and mourning, the tone had shifted to scandalized excitement:

> John W. Parsons, handsome 37-year-old rocket scientist killed Tuesday in a chemical explosion, was one of the founders of a weird semi-religious cult that flourished here about 10 years ago...Old police reports yesterday pictured the former Caltech professor as a man who led a double existence—a down-to-earth explosives expert who dabbled in intellectual necromancy. Possibly he was trying to reconcile fundamental human urges with the inhuman, Buck Rogers type of inventions that sprang from his test tubes.

The lavish rhetoric continued across the tabloids:

> Often an enigma to his friends [he] actually led two lives...In one he probed deep into the scientific fields of speed and sound and stratosphere—and in another he sought the cosmos which man has strived throughout the

ages to attain; to weld science and philosophy and religion into a Utopian existence.

Certainly the photographs of Parsons displayed in every West Coast newspaper suggested that he was an unusual-looking rocket scientist. With his rakish moustache and chiselled good looks, he seemed to emanate an aspect not of scientific stolidity but of Mephistophelean allure.

By June 20, three days after his death, a former member of Parsons' "cult" had been tracked down by the newspapers, whose headlines now ran: SLAIN SCIENTIST PRIEST IN BLACK MAGIC CULT and VENTURES INTO BLACK MAGIC BY BLAST VICTIM REVEALED. The articles told how Parsons had been a high priest of one of the "weirdest cults of mystic potions, free love and exotic ritual ever uncovered in the Southland," and boasted of finding a trail, "locked behind the iron turret of death," that "reached back into the darkest night of the Middle Ages." The unnamed former member explained that Parsons had been a follower of the teachings of one Aleister Crowley, a "British witch doctor," and went on to describe the revels that had taken place at Parsons' home in Pasadena. Lines of robed figures had walked through the grounds, carrying torches and chanting pagan poems. "The score or more of followers were about equally divided as to sexes, and included persons of all ages and professions, some of them brilliant scientists." It was even said that a pregnant woman had disrobed and leapt "several times through 'sacred fire' to insure safe delivery of her child." The fragment of a poem which Parsons had written and circulated among his alleged followers ten years earlier was dutifully printed under the headline POETRY OF MADNESS:

> I height Don Quixote, I live on peyote,
> Marihuana, morphine and cocaine,
> I never knew sadness, but only a madness
> That burns at the heart and the brain.

Other members of Parsons' group also began to speak to the newspapers. One man, who described himself as a "tongue-in-cheek visitor," reiterated that members of the group ranged "from plain screwballs and psychos to some really brilliant scientists," and told how "Friday night seemed to be the big night, and they would run 'round in black robes with daggers at their belts." Parsons, he said, had declared that his "real work" was black magic and had transformed the main sitting room of his Millionaire's Row house into "a temple of hedonistic worship."

In the light of these revelations, even the police criminologist's report on Parsons' death was being reevaluated. While the theory that his death was accidental seemed proven by his injuries, it clashed strongly with the reminiscences of his former work colleagues. In the year before his death, Parsons had been involved in what was called a "confidential research program" for the Bermite Powder Company, a local explosives firm. "Parsons was extremely safety-conscious," claimed one of his colleagues at Bermite. "He worked carefully, had a thorough knowledge of his job and was scrupulously neat." If this was the case, how could Parsons have allowed himself to be in a position in which he might drop his chemicals, let alone be accused of mixing them in a tin coffee can? And why had he been manufacturing a chemical that the army had long ago stopped using precisely because of its volatility?

It was now revealed that Parsons had acted as an "expert witness" during the famed Kynette car-bombing trial of 1938, one of the most shocking in Los Angeles history. Was it more than just a cruel irony that he had been killed in an equally deadly explosion? Fulminate of mercury in a trash can would be anathema to a respected scientist like Parsons. Even the police had to admit it was "incongruous." George Santmyers, an engineer who had associated with Parsons at the Bermite company, compared Parsons' supposed method of manufacturing the chemical to a highly trained surgeon operating with dirty hands: "I intimately knew Parsons as an exceptionally cautious and brilliant scientific researcher." Santmyers suggested that "someone else" had handled the chemicals in Parsons' laboratory prior to his death. MYSTERY ANGLE ENTERS SCIENTIST'S BLAST DEATH, read the *Los Angeles Times*.

The specter of foul play was all too easy to conjure up. This was, after all, Los Angeles, the city that Raymond Chandler was in the midst of depicting as a landscape of murders and femme fatales in such hard-boiled crime classics as *Farewell, My Lovely* and *The Long Goodbye*. It was a city in which things had a habit of going bang in the night. Encouraged by this local instinct for the sensational, the story of Jack Parsons was swiftly transforming itself from family tragedy and scientific calamity into a gothic horror story with a dash of film noir.

Four days after the explosion, in an attempt to quiet the churning of the rumor mill, Detective Lieutenant Cecil Burlingame announced that Parsons' curious disposal of the chemicals wasn't sufficient "to warrant us reopening the case," nor was his membership of a religious cult. "His death is listed as an accident," announced Burlingame. "The case is closed as far as we're concerned."

However, this official judgment did little to disperse the increasing hubbub of gossip and conjecture that continued to swirl through Los Angeles and Pasadena among Parsons' friends and former acquaintances. Soon Parsons' bizarre story was spreading into the national magazines. The following month's issue of *People Today* published a profile of Parsons under the title, L.A.'S LUST CULT:

Rich rock-ribbed Pasadena, famed for its roses, the California Institute of Technology and as a retirement haven for Eastern millionaires, looks like the last place a black magic cult dedicated to sex would thrive. Nevertheless the Church of Thelema (the name means "will"), a cult practising sexual perversion, has been making converts of all ages, sexes, there since 1940. Among the believers: many prominent residents of the Pasadena-Los Angeles area; at least one member of Hollywood's movie colony. The existence of the cult ... was only proven this June by the "accidental" death of high priest John W. Parsons.

Speculation continued to mount. Some of Parsons' former lodgers suggested that he had been depressed for some time and that his death had been a spectacular suicide. Others who had gotten to know him through his love of science fiction magazines imagined, only half-jokingly, that he had been trying to conjure up a homunculus, a magical creature created by the alchemists of yore, but that the magical working had gone wrong and by mistake he had summoned a fire demon that had consumed him. The scientific establishment, however, remained resolutely silent on the matter. Despite Parsons' obvious prominence in their ranks, any comments they made were curiously oblique. When Aerojet's secretary-treasurer was asked about him, he described Parsons simply as "a loner" who "liked to wander." Parsons' character seemed to be changing by the day, becoming less real, more exaggerated. Indeed, in the eyes of the public and the press, John Whiteside Parsons had gone from being a young genius dead before his time to the most overworked, hackneyed, science fiction cliché of them all—the mad scientist.

The fantastical, tragic, and largely unknown story of John Whiteside Parsons is one of the most intriguing tales to be found in the annals of modern science. His life brings together the seemingly disparate worlds of rocketry, science fiction, and the occult. But for Parsons there was never any contradiction in these subjects.

For us, the rocket scientist exemplifies intellectual complexity. The phrase "it's not rocket science" instantly places rocketry as the ceiling on cerebral understanding. When Jack Parsons began shooting off homemade rockets in his backyard in the rockets, was not only not a science; it had not even been coined as a word yet.

During the first decades of the twentieth century the world was in awe of aviation. Since the Wright brothers' historic twelve-second flight in 1903, pilots had swiftly become modern-day heroes. By the time Charles Lindbergh flew solo across the Atlantic in 1927, airplane manufacture had become the boom industry of the era. The same could not be said of rockets. Despite having been used in fireworks and primitive weapons for over a thousand years, these often complex machines had never been comprehensively studied. No universities taught rocketry courses, and there were no government grants allotted to rocketry research. In established scientific circles, rockets were synonymous with the ridiculous, the far-fetched, the lunatic, as much a euphemism for "foolish" as *rocket scientist* is now a byword for "genius."

Ironically it was in the United States that both public and professional opinion was especially hostile. A widely used textbook on astronomy, released in 1933, patronizingly claimed to understand the appeal of rockets and its attendant goal of space travel, but decreed that there was "no hope" that such wishes could ever be realized; "only those who are unfamiliar with the physical factors involved believe that such adventures will ever pass beyond the realms of fancy."

It was precisely these realms of fancy that inspired amateur enthusiasts across the globe to begin experimenting. As the historian of astronautics Frank H. Winter has written, "Science fiction was, in the beginning, an inseparable and formidable factor in fomenting ideas about spaceflight," even though the genre was derided by the general public and the press alike as a juvenile and inconsequential form of literature. Inspired by these futuristic stories, amateur rocketeers formed space travel clubs and intended to develop rockets not for entertainment or for weapons, but for the cause of space exploration. Holding up the science fiction magazines as their scriptures, enthusiasts from all walks of life constructed small, primitive rockets, hoping to progress toward their far-off goal (though more often than not these rockets blew up on takeoff or exploded in midair).

Space travel was a brave dream, largely because it was sought in the face of so much public and professional hostility. Even as late as 1941, one rocketry enthusiast was mocked in Congress as "a crackpot with mental delusions that we can travel to the moon!" at which the entire House of Representatives roared with laughter. Such mockery was unsurprising. We like to think of our sciences as cumulative enterprises, incorporating centuries of thought within their practice and a pantheon of innovators stretching back to antiquity. Rocketry was not like this. Up until the twentieth century it was a threadbare discipline, with few maxims and fewer heroes, lacking the deep theoretical and experimental foundations on which any science is based. In rocketry these fundamentals would grow out of the amateurs' dreams and "delusions." "Not the public will, but private fanaticism drove men to the moon," declared the sociologist William Bainbridge. Those interested in rockets were as much obsessive visionaries as technical geniuses.

Jack Parsons was just such a figure, living on the cusp between an old world in which the very idea of space travel was a scientific absurdity and a new world in which it would become scientific fact. It was this new world which, despite his lack of a university degree or professional scientific qualifications, he would help to create. Along with his motley band of experimenters, disparagingly known as the "Suicide Squad," he revolutionized the public and academic perception of rocketry, transforming it from an object of ridicule into a viable science. In the process Parsons invented a radically new kind of fuel, the descendants of which are still used in the space shuttle to this day, and helped found the Jet Propulsion Laboratory at Caltech, which has since become the world's preeminent institution for the exploration of the solar system. In many respects the United States' path to the moon landings began with him. In the words of the great scientist Theodore von Kármán, after the work of Parsons and his partners, "a new age was born."

Parsons himself was born into an age in which perceptions of the world were changing on a daily basis. Laws that had been set in stone were swiftly crumbling under the advance of science. In 1916 Einstein published his *The Foundation of the General Theory of Relativity;* in 1927 the big bang theory of the universe was introduced; and in 1930 the planet Pluto was first discovered. The age was illuminating, confusing, and frightening. When Niels Bohr, one of the greatest interpreters of the new science of quantum theory, stated, "If you aren't confused by quantum physics, then you haven't understood it," he seemed to be ushering in a mysterious new era of chaos and absurdity. Naturally there were backlashes against this scientific revolution. The old known world was not going to give up easily. The Scopes trial of 1925 in which a biology teacher from a Tennessee public school was convicted of teaching Darwinism in the classroom was a case in point.

Los Angeles, where Parsons spent the early years of his life, was a metropolis perfectly in tune with this perplexing time. During Parsons' youth, evangelists like Aimee Semple McPherson might be heard performing exorcisms over the new medium of radio, broadcasting to hundreds of thousands of people, while Albert Einstein, harbinger of the scientific age, attended a séance hosted by a dubious Polish count. Igor Stravinsky, the most famed composer of the era, ended up in the city providing music for Walt Disney's *Fantasia,* while the prominent astronomer Edwin Hubble could be found dining with the mime Harpo Marx. The author William Faulkner was reduced to rewriting B movie film scripts, while the social campaigner and writer Upton Sinclair was arrested for reading the First Amendment of the United States Constitution (the right to freedom of expression) in public. All was topsy-turvy, nothing more so than Los Angeles itself, an Ozymandian kingdom built on a desert that had been transformed by the wonders of engineering into fertile land.

Jack Parsons' life exemplifies this place and age of flux and uncertainty. When I first came across his story, I was amazed by its bizarre contradictions. First and foremost was the fact that one of America's pioneering rocket scientists was also a devout occultist, fascinated by magic and the supernatural. But as I delved deeper into his life, the strange mixture of science and magic was only one of many incongruities that appeared. How could a college dropout find himself, at the age of twenty-six, a government-funded rocket scientist? How did he come to live a bohemian, free love lifestyle amidst the social strictures of the 1930s and 1940s? Why did he exude "an aura of inherited wealth" and yet have to scrounge for money for his rocketry experiments? And what was he doing appearing in both scientific journals and science fiction stories? I have long been fascinated with Los Angeles as a crucible from out of which the world's trends erupt. Parsons seemed to embody the city's tumultuous character.

I soon found that Parsons' story, besides being a guide to his times, also helped elucidate the process of scientific discovery. Most scientific research is based on past achievements in the field, its accomplishments recounted in textbooks and taught in lecture halls. But when there are no textbooks to read, as in the case of rocketry, where does one turn for inspiration? Parsons' story reassures us that at the heart of all scientific advances is the imagination—that what we perceive as perverse eccentricities can be the key to important breakthroughs.

In Parsons' case, his obsession with magic placed him in a long line of scientists stretching back to antiquity who have explored the occult. These include Robert Boyle, the celebrated seventeenth century chemist; Dr. John Dee, the court astronomer to Elizabeth I, and most famous of all, the father of the Age of Reason himself, Sir Isaac Newton.

While Newton is largely responsible for the scientific enlightenment that swept away the common belief in magic and mysticism, he also immersed himself in these very same practices.

Newton did not call himself a scientist (the term was not coined until 1834). He was a natural philosopher and an adventurer of the intellect. Newton's most famous work, the *Principia Mathematica,* advances a highly sophisticated and complex description of the workings of the universe; but he was also fascinated by alchemy—the ancient precursor of the science of chemistry—in particular its quest for the philosophers' stone, which was said to have the power to transform any base metal into gold. Such was Newton's enthusiasm for this subject that the distinguished economist and Newton scholar, John Maynard Keynes, described him as "Copernicus and Faustus as one." "Newton was not the first of the age of reason," wrote Keynes. "He was the last of the magicians, the last of the Babylonians and Sumerians, the last great mind which looked out on the visible and intellectual world with the same eyes as those who began to build our intellectual inheritance rather less than 10,000 years ago."

Like his illustrious predecessor, Parsons did not see the two disciplines of science and magic as contradictory. Writing three years before his death, Parsons stated with a certain sober detachment, "It has seemed to me that if I had the genius to found the jet propulsion field in the US, and found a multimillion dollar corporation and a world renowned research laboratory, then I should also be able to apply this genius in the magical field."

He treated magic and rocketry as different sides of the same coin: Both had been disparaged, both derided as impossible, but because of this both presented themselves as challenges to be conquered. Rocketry postulated that we should no longer see ourselves as creatures chained to the earth but as beings capable of exploring the universe. Similarly, magic suggested there were unseen metaphysical worlds that existed and could be explored with the right knowledge. Both rocketry and magic were rebellions against the very limits of human existence; in striving for one challenge he could not help but strive for the other.

———

I first encountered Jack Parsons as little more than a footnote in a history of rocketry—appropriately, since he has been relegated to the sidelines of that history ever since his death. One reason for his marginalization seems to have been the embarrassment his unorthodox lifestyle caused his academic successors; another is his elusiveness. When Parsons died, he left no heir, and many of his letters and documents have since been lost or destroyed. Also lost are many of the minutes and papers concerning the Jet Propulsion Laboratory's earliest days, and many of the central characters featured in this book—Frank Malina, his rocketry colleague; Ed Forman, his closest childhood friend; Theodore von Kármán, the scientific great; and L. Ron Hubbard, the fantasy writer—are dead.

Nevertheless, through his first wife, the late Helen Parsons Smith, through the memories of those who knew him in Pasadena and at Caltech, through the archives kept by the Jet Propulsion Laboratory, the Ordo Templi Orientis—the occult society of which he was a member—and the painstaking ministrations of a few eager archivists and enthusiasts, I managed to learn a great deal about Jack Parsons. This book is intended to show him in all his peculiar glory for the first time. In writing it I hope to free Parsons from both establishment censure and mystical titillation, from the footnote, and from the "mad scientist" tag.

Upon first looking into Parsons' story, I found him a fearsome figure, dour and surrounded by occult dogma. But the more his friends and former acquaintances talked about him, the more the human being came into view. Those whom I spoke to recalled him with fondness and an amused exasperation at his impetuosity. Each told me of his charisma, his brilliance, his enthusiasm, but also of a man whose total dedication to his science and way of life could leave him indifferent to others' emotions, aloof to the real world. He seemed to create for himself various personae—the literary dilettante, the rocket scientist, the magician—which may be why he remained something of a mystery to even those who knew him well. When his rocketry work was not recognized or when events contradicted his self-created myth, he was prone to deep depressions and extreme mood swings.

But at the heart of his character was an essential optimism, a confidence that if he believed in an endeavor enough, he would eventually gain the prize. Parsons was by no means an innocent, but he possessed a child's capacity to believe, a naiveté, as well as a love of experimentation. It was this mindset in particular that allowed him to break scientific barriers previously thought to be indestructible.

Ultimately his insouciance and his otherworldliness would lead to his scientific downfall. The enthusiasms and complications of his private life would overpower him and be ruthlessly exploited by others. He would retreat further into his magic as it became the only world he could control. The man who had done so much to establish the science of rocketry in America would end his life making special effects for Hollywood film companies.

Nevertheless, his willingness to believe in magic, to be inspired by science fiction, to dare to challenge the scientific establishment, humanizes what has since become a strangely antiseptic and colorless discipline. Like many scientific mavericks, Parsons was discarded by the establishment once he had served his purpose. But in the short time he existed he represented a character that is less and less prevalent in the world of science today: the wide-eyed dreamer, the visionary scientist. In his wish to push the world into the future, he can be seen as the brother of the American pioneer, or his modern-day counterpart, the space explorer of science fiction. His life suggested that it is sometimes by going in the irrational and unknown direction that great leaps forward can be made. Jack Parsons' story is that of the traveler seeking a brave new world.

1
. . . .

PARADISE

*The paradox implausible, the illusion that
must be seen to be believed.*

—RAY BRADBURY,
Los Angeles Is the Best Place in America

In December 1913 Ruth and Marvel Parsons left the ice and
snow of the East for what they hoped would be a new future.
Woodrow Wilson had recently been declared the twenty-eighth
president, and while all Europe watched the increasing tensions
in the Balkans, many Americans were turning their backs on the
Old World and looking towards the warm promise of their very
own West.

Ever since gold had been discovered in California in 1848,
thousands upon thousands of people had poured towards the
Pacific Coast, flooding a state which up until then had had a
population of barely 18,000. The alchemical surge of the gold
rush brought not just prospectors but their attendants—the
thief, the cardsharp, and the minister, the last intent on convert-
ing the hordes set free from the laws and moral codes of the
East. It was not an easy task. California, declared one Methodist
preacher, was "the hardest country in the world in which to get

sinners converted"; indeed, "to get a man to look through a lump of gold into eternity" was nigh impossible.

By 1913 most of the gold had disappeared, but the transmutative effect of the rush survived. The promise of a golden *life* was now the prize. Agriculture had surpassed mining as the state's biggest industry, and California was transformed into the Garden of America, creating for itself a reputation as a land of orange groves, vineyards, flowers, and sunshine. A health rush succeeded the gold one, as doctors who regularly prescribed a change of climate to deal with a long list of complaints and disorders now suggested California as the ultimate cure. The state would always retain its symbolic connection with that most persuasive of American myths, the pursuit of happiness.

The young couple now traveling by railroad through the freezing winter had married just the previous year in the bride's hometown of Springfield, Massachusetts. Ruth Virginia Whiteside, the only child of Walter Hunter Whiteside and Carrie Virginia Kendell Whiteside, was twenty-two years old when she married. Doted on by her parents, she had lived a sheltered life, growing up in a wealthy manufacturing family in Chicago. Her father had been hugely successful as the president of the Allis Chalmers farm equipment company before taking over the reins of the Stevens-Duryea automobile corporation in Springfield. There Ruth met Marvel H. Parsons, a man's man two years her senior, who loved the great outdoors and whose family had founded the town of Springfield in the early seventeenth century. His unusual first name had come from his mother, Addie M. Marvel, but he was known to all by the less awkward name of "Tad" or "Teddy." The marriage had seemed a good match, a consolidation of middle-class fortunes: Marvel's father was a real estate developer who had codeveloped the Colony Hills neighborhood just outside Springfield. He was also president of the Eastern States Refrigeration Company, which owned warehouses extending along the Grand Junction Wharves in

Boston. Yet for all its financial sense, Ruth and Marvel's union was ill-starred.

Within less than a year of the wedding, Ruth gave birth to their first child. It was stillborn. The young couple was devastated, particularly Ruth. With her health fragile and their home in Springfield clouded by tragedy, a move away from the East was thought best. It did not take long to choose a destination. Nowhere were the surroundings more propitious, the opportunities more abundant, or the boosters more feverish than in Los Angeles, the ecstatic beating heart of the Land of Sunshine.

It had not always been so. Founded as a Mexican colony in 1781, Los Angeles was a stagnant pueblo for nearly a century. By 1850 the city housed little more than 8,000 inhabitants and was known as the "Queen of the Cow Counties" from its role as the trading center of the southern Californian beef industry. Under American occupation it had transformed itself from a sleepy settlement into a violent border town. A motley assortment of "cowboys, gamblers, bandits and desperadoes" drawn both by the cattle and the possibility of gold ensured that one murder was committed for every day of the year. The Reverend James Woods, a visiting missionary, was shocked by the lawlessness, drunkenness, and low regard for human life he saw. "The name of this city is in Spanish the city of angels," he wrote in his diary, "but with much more truth might it be called at present the city of Demons."

But in the decades that followed, unprecedented floods and drought saw the cattle industry falter. With the construction of the Southern Pacific Railroad and the city's shift from cow town to farming center, more and more well-heeled immigrants began to arrive. By the end of the nineteenth century the hell that the Reverend Woods had set eyes on had been transformed into its exact opposite.

"We have a tradition," wrote one Californian journalist, "which points, indeed, to the vicinity of Los Angeles, the City of the Angels, as the site of the very Paradise, and the graves are actually shown of Adam and Eve, father and mother of man and (through some error, doubtless, since it is disputed that he died) of the serpent also."

Boosterism on the biblical scale became common and reinforced what the gold and health rushes had already proven: that here was a place to redeem oneself, to return to the garden before the Fall, to sever all connections with the past and, hopefully, to make a wondrous new beginning.

In 1910, Los Angeles had 319,198 residents, a sixfold increase from twenty years before. But that growth would be dwarfed by what was to follow. When Ruth and Marvel arrived three years later, William Mulholland, the city's chief engineer, had just opened the first aqueduct into the desert city. As the water poured through it, ensuring the city's urban destiny, Mulholland spoke as if he had co-opted divinity into his scheme. "There it is," he proclaimed, "take it." And the people did. More and more took it each year. The Californian dream was the belief that fantasy just might be made into reality, the dream that people, like the resources of California itself, could be tapped and transformed from barren disappointments into verdant successes.

Los Angeles was now a sprawling, bustling city, spreading over some sixty-two square miles and rapidly incorporating the surrounding communities, most noticeably Hollywood, which had already begun attracting film companies with its climate fit for year-round filming. Along with real estate, cars, and shipping, filmmaking would soon become one of the city's largest industries. Los Angeles architecture was a patchwork of styles, combining elements of the Spanish mission designs of yore with the ranch house of the American Midwest. The garden bungalow became the preferred form of housing, and the automobile was swiftly becoming a key component of city life, as ubiquitous as the electric streetcars.

The Parsons settled into a house at 2375 Scarf Street, just south of downtown Los Angeles. The munificence of their respective families had helped pay for the couple's journey westwards, but now they had to fend for themselves. Marvel found himself a modest job at the P. A. English Motor Car Company on South Grand, selling auto accessories to the ever increasing number of car owners. The new metropolis entranced him. In the words of the Californian critic Carey McWilliams, Los Angeles was not so much an urban landscape as "a great circus without a tent." Inhabitants came not only from across the United States but from China, Japan, the Philippines, India, and Mexico, providing the majority of the farm labor force and bringing with them many of their customs and religions.

Attire on the streets of the city ranged from straw hats to fur coats. Electric signs blazed everywhere; "clairvoyants, palm readers, Hindu frauds, crazy cults, fake healers, Chinese doctors" all plied their trade. In 1906 over 50 percent of Los Angeles' population may have been Protestant, reflecting the number of transplants from the midwestern states, but a whole new breed of radical metaphysical religions, such as Christian Science, New Thought, and Theosophy, had begun to take root alongside the mainstream beliefs. Confucianism, which had arrived via Chinese immigrants, began to seep its way into the sermons of some of the more liberal Protestant churches. Spiritualism found proponents of its creed of mystical development and séances, especially in the Hollywood film community where it was now becoming something of a craze. Secular utopian communes were also springing up outside the city, most notably the short-lived Socialist community of Llano del Rio which at its peak had over 1,000 self-sufficient men, women, and children farming 10,000 acres of land.

Despite the vast number of religious groups and the fact that the Anti-Saloon League of California had suppressed virtually every drinking establishment in Los Angeles by 1910, organized vice was rife, and many of the police force were on the take, foreshadowing the corruption that would be another of the city's defining features. Brothels could frequently be found on the same street as churches, and although evangelists did their best to paint a veneer of moral rectitude over the immoral proclivities of the city, they instead imbued it with a quality of schizophrenia.

The Parsons decided to celebrate their arrival in town by trying for another child, and this time there was to be no heartache. Almost ten months after his parents set foot in Los Angeles, Marvel Whiteside Parsons was born at the Good Samaritan Hospital on October 2, 1914. As his father had always gone by the nickname Tad or Teddy, so the new addition to the family was also helped out of his unusual moniker; his parents called him Jack.

The new family moved into a bigger house at 2401 Romeo Street, just off the long stretch of Wilshire Boulevard that ran to the northwest of the city center. But rather than solidifying the marriage, the arrival of little Jack heralded its end. Los Angeles lacked many of the social strictures of staid Massachusetts, and Marvel Parsons was pursuing the vices of the city with reckless abandon. In the months before Jack's birth and in the immediate weeks after it, he made frequent visits to a prostitute. Whether he was caught in flagrante delicto or whether he admitted his wrongdoing in a fit of guilt, we have to imagine; surviving letters do not say. However, by January 1915, two and a half years after they were married, Ruth had forced Marvel to move out of the house on ill-named Romeo Street.

It was a bitter split. Marvel Parsons continued to live and work in Los Angeles and write Ruth long, pained letters in which he begged for forgiveness. He wanted to return to the house but was afraid "of being shot or scaring [her] to death." His letters suggest the frantic anger which Ruth now felt. Having lost her first child, abandoned her hometown, and given birth to a son, she had been rewarded with Marvel's unfaithfulness. If Ruth had been a demure and fragile New Englander up until now, her husband's infidelity demonstrated how ferocious she could be.

Marvel tried desperately to soothe Ruth's anger to persuade her that his act had meant nothing. "Ruth I may be very brutal but I think you are very foolish to have the thoughts you have of the other woman ... Do you think I love that sort of woman ... Love—you are crazy to think I love her or anyone else but you. Haven't you learned that it is anything except love that let's [sic] a man stay with a prostitute."

He also tried to convince Ruth that she was being unreasonable, impressing upon her the fact that they were living in a new, less restrictive age. "I think Ruth to be honest that I was brought up as the average boy is brought up while you were brought up as only one woman in a thousand. Your ideals and standards are not of the world today. They are beautiful, some day they may come true for the world, but not in our generation." But Ruth was not to be placated, and she must have ignored Marvel's arguments: No letters from her exist, while his imply that he is meeting a stony silence, even when he pleads with her to be able to see "Little Jack."

By March 1915 Ruth had initiated divorce proceedings. Censure of divorce had lessened somewhat since the proscriptive Victorian era, and Los Angeles in particular had one of the highest divorce rates in the country, with one in six marriages ending in the courts. Marvel, finally realizing that he had no chance of winning Ruth back, meekly asked her not to name adultery as the cause. Ruth ignored him and, after the divorce was finalized, cut off all communication. Publicly named as an adulterer and unable to see his child, Marvel chose to return home to Massachusetts. He had moved to the west for his wife's sake. Now she wanted nothing to do with him. He continued to write to her sporadically. "Do you think it is quite fair," he writes in one of his letters, "not to write me once in awhile how the boy is?" Again there was no reply. "Pretty hard to sit here," he says, defeated, "and think that my own son is not being taught to say 'papa'."

Indeed, Jack would never truly know his father, and Ruth Parsons made sure that no reference was ever made to his first name, Marvel. Her son was to be referred to as "John White-side Parsons" on all official documents.

We can guess at the depth of Parsons' reaction to this loss because he later wrote about it—something he rarely did. His father's absence was a central theme of a brief autobiography he wrote at a time of extreme emotional despair in his midthirties. The manuscript, written in the second person ("Your father separated from your mother in order that [you] might grow up with a hatred of authority"), is part psychoanalytic autobiography, part self-mythologizing reinvention of his life. It is by turns painfully honest and disconcertingly impassive and suggests that his childhood relationship with his mother became especially close to compensate for his father's loss. Indeed, the search for a father figure would occupy Parsons throughout his life.

Nevertheless, his mother was not to be the sole influence on his early years. Shortly after hearing of their son-in-law's adultery and their daughter's insistence that there would be no reconciliation, Walter and Carrie Whiteside decided that since they were nearing retirement age and were wealthy to boot, they would move west to live with their only daughter and grandchild. The house on Romeo Street was abandoned, and the Whitesides bought a home in a suburb of Los Angeles that was increasingly attracting the wealthiest and most sophisticated members of society to its hallowed ground—Pasadena.

Following the bitterly cold winter of 1872, Dr. Thomas Elliott of Indianapolis decided that he and his friends had suffered long enough the Midwest's inhospitable climate. In order to escape the colds, coughs, and chills that had troubled them and their families for so long and to get "where life was easy," he formed the California Colony of Indiana. Surveyors were sent out to find suitable land and within months they had it—four thousand acres of the "fairest portion of California" at the western head of the San Gabriel Valley. The parcel was beautifully situated. Sheltered by the mile-high San Gabriel Mountains, it enjoyed perpetual sunshine, contained an abundance of colorful local flora, and was conveniently located just ten miles from the growing urban center of Los Angeles. Soon the area was subdivided; cottages were built and orange groves planted. By 1875 the Indiana colony had acquired a post office and named themselves Pasadena (the Chippewa name for "valley"), and ten years later Pasadena was linked by rail with Los Angeles and Chicago. "Pullman emigrants" rolled into the town, many coming, like the original colonists, to escape chronic ailments such as tuberculosis, for which the dry Pasadena air was famed as a cure. Within ten years it had become the premier resort town in the country.

Mount Lowe and Mount Wilson both hunched over the city to the north. Those who climbed to their pine-clad peaks and cast their gaze back down below would have been charmed by the prospect. Pasadena looked like a sea of green trees through which numerous white church spires protruded. Giant hotels could be seen resting amidst the orange groves, enticing wealthy tourists from the East and Midwest to lengthen their visits and become citizens of Pasadena.

"It is the land of the afternoon," wrote resident Charles Frederick Holder. "People live out of doors and have an inherent love of flowers." While the rest of the country froze, Pasadena gloated over its natural abundance at the New Year's Day Festival, better known as the Tournament of the Roses. Since 1890 the city had honored its floral glut in a truly Arcadian communal boast. Foot races were run, games were played, and chariots were raced. There was even a jousting match in which horsemen with lances tried to spear three rings hanging at thirty-foot distances. But the day's centerpiece was the parade of flower-bedecked carriages that wove through the city's streets, ridden by Pasadena's beaming beauties who threw flowers as they went.

The city was swiftly becoming a Mecca for visiting architects as new and uniquely Californian designs were constructed in the city. Henry and Charles Greene almost single-handedly created the Craftsman style with the large wooden bungalows they built for their Pasadena clients. Calling upon Swiss and Japanese influences and using materials gathered from the surrounding wilderness, they designed houses that were poems of wood, texture, and light, featuring open beams, skylights, stained glass windows, and low slung eaves, fit for the fancy of any American who sought to establish a gilded frontier lifestyle.

But while residents basked in the languor of their pioneer daydreams, the city still retained something of the intellectual energy and progressive spirit of its midwestern Protestant origins. When the opera or symphony played in Los Angeles, the Pacific Electric Railway ran special cars to and from Pasadena. New schools and centers of learning were constantly being built, and groups such as the Chautauqua Literary and Scientific Circle and the Social Purity Club were swift to make their presence felt with lectures and dances. Astronomers began studying the heavens from the Mount Wilson Observatory, built by the astronomer George Ellery Hale; and the local technical college, Throop, was slowly undergoing its transformation into the California Institute of Technology. It was not long before Pasadena became known as a "western clearing house for Eastern Genius."

By the turn of the century, Pasadena had already been visited by two presidents, and residents such as Jason and Owen Brown, the sons of the famed abolitionist John Brown, attested to the city's moral seriousness and political leanings. When a third president, Theodore Roosevelt, visited in 1903, Pasadena's importance was assured. The editor of the *Pasadena Daily News* spoke of the city in 1907 as embodying all that is "beautiful, clean, cultured, moral and aesthetic." Such traits were not gained by accident. While the flood of immigrants saw Los Angeles sprawl, Pasadena was adamant about rebuffing the less appealing by-products of urban growth. The Pasadena Board of Trade, run by many of the wealthiest residents, consistently voted against encroachments of factories or large-scale business enterprises. "We do not bid for factories," said a contemptuous president of the board, D. W. Coolidge, "but lay special stress on our superior location, climate, civic improvement, churches and schools as making the most desirable place of abode." In 1906 only an estimated 10 percent of the population were classed as "laborers and artisans." By 1920 Pasadena had the highest per capita income of any city of its size in the country; and by 1930 the city, whose population was now over 76,000, could still claim domestic servants as its largest labor force.

In comparison to the conservative, antiunion governance that held sway ten miles away in Los Angeles, Pasadena harbored pockets of left-wing thought. While the oligarchs of Los Angeles were fighting pitched battles against labor in order to attract businesses to the West Coast, Pasadena—which had no want, or need, for businesses—paradoxically set itself up as one of the friendliest municipalities to the unions. Open shop bands were often thrown out of the Tournament of Roses parade, and the American Civil Liberties Union was permitted to speak in Pasadena in spite of fervent opposition from the strictly conservative American Legion and Better America Federation. Indeed, Pasadena would gain its full Socialist credentials when Upton Sinclair, author of such jeremiads against big business as *The Jungle* and *Oil!* moved to the town (even though his uncommon interest in the working man saw him shunned by the city's uppermost echelons). Pasadena was enlightened, moral, aesthetic, and rich. If people traveled to Los Angeles to achieve their dreams, they moved to Pasadena when they had attained them. It was the VIP enclosure of Paradise.

If Pasadena was the jewel of Southern California, then the jewel of Pasadena was Orange Grove Avenue. Unlike the grid system that shaped the rest of the city, Orange Grove had been laid out at a three-degree angle from true north to preserve some of the area's native oaks, which now stood obstinately in the middle of the road. By the time Parsons' family moved to Pasadena in 1916, some fifty-two millionaires populated the

one-and-a-half-mile-long avenue. They included Lamon Van-
derburg Harkness of New York City, one of the richest men in
the world because of his Standard Oil Company, which had
been recently dissolved by Supreme Court decree. Arthur Fleming, the Canadian-born logging magnate and philanthropist,
lived in the first Craftsman-style house to be built in Pasadena.
Chicago chewing gum millionaire William J. Wrigley lived in an
Italianate mansion at the top of the avenue, while Dr. Adalbert
Feynes, a famed entomologist, resided in an Algerian-style
palace not far away. The St. Louis beer millionaire, Adolphus
Busch, had created a giant stone mansion overlooking his wondrous gardens, and the black-clad widow of assassinated president James A. Garfield also lived on Orange Grove. Behind
ivy-clad walls, manicured hedgerows, and twelve-foot-high pil-
lared gates were vast estates with swimming pools and tennis
courts, driveways encircled with roses and flowering vines of
perpetual summer. Footmen and even carriages were sometimes
seen on the road. And where did these great and good meet? At
the northernmost end of the avenue, where the Valley Hunt
Club acted as the exclusive preserve of Pasadena's high society.

If any one street was responsible for the civilizing of California in the minds of the New England Brahmins, then Orange
Grove Avenue was it, for it would have taken a stubbornness
even greater than that displayed in the Episcopalian East not to
have been impressed by the sheer mellifluous quality of Orange
Grove. Languor and energy, rusticity and sophistication, American nature and European art mingled sweetly throughout. The
avenue made even the palm trees look dignified. The year that
the Parsons arrived on Orange Grove was the year the *Los Angeles Times* named it "the most beautiful residence street in the
world."

Not one to shy away from ostentation, Walter Whiteside
purchased a giant Italian-style villa at 537 Orange Grove Avenue so that his small, multi-generational family could rival any
of the more conventional dynasties of Pasadena. Set back from

the road amidst an acre and a half of cosseted foliage, the house met the visitor with a facade of pristine stucco, shaded windows, and sculpted arches. Within the cool walls the family of four shared some twenty rooms with their two English servants. What's more, this mansion sat right next door to the Valley Hunt Club.

Jack Parsons spent almost all of his childhood surrounded by this prodigious wealth. His earliest memories would have been of an exotic palace that seemed his alone, with attentive servants compliant to his every need. As the sole child in the house, he was given strict lessons in manners and treated as the heir apparent to the Whiteside family by his doting grandfather. As for Ruth, blue-blooded Pasadena suited her much better than downtown Los Angeles, and she swiftly entered into the social whirl that occupied Pasadena's elite—chamber concerts with the Pasadena Music and Art Association, lectures at the Twilight Club, theater at the Pasadena Playhouse, golf and tennis tournaments at the Valley Hunt Club, and maybe the odd trip to the polo fields a few miles away. On one occasion the world renowned Austrian opera singer, Madame Schumann-Heink—better known in the popular press as "The Heink"—sang a private recital for the family, with young Jack sitting on her ample knee.

Parsons' neighborhood was no less fantastical than his home. Crenelated French chateaux with faux-arrow slits stood side by side with the domes and crescent moons of Moorish palaces, while the vast Craftsman-style bungalows conjured up pictures of the Orient with their sloping beams and precise lines. To the south of Parsons' house lay the Busch Gardens, consisting of thirty acres of manicured lawns and exquisite floral pageantry. Plants from all over the world bordered the rolling lawns, and over fourteen miles of walkways wound their way through the gardens, serving thousands of tourists a year (not to mention numerous Hollywood film companies). The miniature buildings and statues scattered throughout the grounds, especially designed to delight the younger visitors, added a touch of magic. Under a darkening bower could be found a small cottage plucked straight from "Hansel and Gretel," while closer inspection of a fountain revealed a horde of tiny terra-cotta fairies. In such an idyllic setting a young child could easily lose himself in imagining that such tales were true and that such creatures existed, if not here, then where?

And if the pampered and irrigated environs of the Busch gardens felt too genteel, a real wilderness could be found backing up to it. The Arroyo Seco (dry stream) cut deep into the landscape along Pasadena's western edge. Here facets of the old frontier still survived. Chaparral covered slopes, and steep rocky sides formed a natural playground for young and old alike. The valley floor was thick with sycamores and tangled thickets of wild grapes. Rabbit and deer could be hunted among the spruce, oak, and bay, and the town's children made camps, fired BB guns and let off firecrackers. A touch of surrealism was added by the ostrich farm positioned at the valley's southernmost end. Part Huckleberry Finn playground, part never-never land, Pasadena provided the perfect landscape for the imaginative child, with Orange Grove its most blissfully secluded centerpiece. It was little surprise that Parsons grew up unconstrained by reality. Throughout his entire life he would never feel more at home, or more at ease, than when living on this fanciful street.

2

MOON CHILD

Mankind will not remain bound to the earth forever.

—from the obelisk of the Russian space pioneer
Konstantin Tsiolkovsky

As a boy, Parsons suffered from two of the hazards of being a single child: He was spoilt, and he was solitary. He had few friends, a fact which he would later see as a great boon in developing what he called "the necessary background of literature and scholarship." Without television and with Pasadena shunning the movie halls that inundated Los Angeles, Parsons read voraciously. His taste tended towards classical tales of romance and fantasy, and he devoured the Arthurian legends, the *Arabian Nights,* the legends of Greek and Norse mythology, and stories of ancient battles. "When he was a youngster he used to read about King Arthur," a friend from later in his life recalled. "It was a dream of his as a child to belong to a group of men who were doing something noble and wonderful. And he also wanted to go to the moon."

This last dream was provided for him by Jules Verne's classic fantasy of 1865, *De la Terre à la Lune (From the Earth to the Moon).* Verne tells the story of a group of demobilized American soldiers, members of an old artillery company known as the "Gun Club," who, in seeking an outlet for their frustrated aggression and some use for their ballistic talents, design a plan to literally blast themselves into outer space. By inventing a new explosive powder and constructing a nine-hundred-foot long cannon, they shoot themselves free of the earth's atmosphere and enter into orbit above the moon. What is most striking about the book is its scientific realism. No scientist of the era would have said space flight was possible, but Verne described it in great detail, using only the technical knowledge available at the time. His talk of velocities and materials, the minutiae of technical method, gave the fantasy that most enticing ingredient—plausibility. To the young Parsons space travel must have seemed only steps away.

Verne's story sent Parsons to the pages of pulp magazines for other tales featuring space travel or scientific themes. The pulps—so named because of the inexpensive paper on which they were printed—had been part of a magazine publishing revolution in the 1880s. Magazines had once had relatively small circulations and had been aimed at the upper middle classes, but a universal rise in literacy had created a craze for enjoyable and affordable reads. With gaudy covers and sensational stories, the pulps were the television of their day. The *Argosy,* a 192-page weekly which featured adventure fiction by such writers as Edgar Rice Burroughs, the inventor of Tarzan, had a readership by the turn of the century of some 700,000. Along with its numerous competitors, the *Argosy* delivered a weekly diet of shoot-outs, monsters, and murders, and while the majority of stories might not have been very technically competent, the fans' enthusiasm for them was unmistakable.

By the time Parsons could read, the pulps had diversified into hundreds of specific genres. *Black Mask,* founded by the literary polymath H. L. Mencken, specialized in crime stories and would later feature the writings of hard-boiled noir writers

Raymond Chandler and Dashiel Hammett. *Ace-High Western Stories* was dedicated to cowboy yarns and gunfights, and *Weird Tales*, which included the macabre fiction of H. P. Lovecraft, spun sword and sorcery and horror yarns. There was as yet no publication devoted to stories like Verne's scientific romances, so Parsons would have had to content himself with the occasional tales of the future and new technologies that would appear in the other pulps.

However, when Parsons was twelve years old, a new magazine for boys entered the market. With a garish yellow cover picturing a red-and-white planet and what appeared to be ice-skating aliens, *Amazing Stories* became the first magazine devoted solely to space-age fantasies. Its editor was Hugo Gernsback, a writer and inventor from Luxembourg who harbored deep utopian leanings for his populist publication; the magazine's motto was "Extravagant Fiction Today...Cold Fact Tomorrow." *Amazing* was to publish stories of what Gernsback named "scientifiction." "By 'scientifiction'," he wrote in his first editorial, "I mean the Jules Verne, H. G. Wells, and Edgar Allen Poe type of story—a charming romance intermingled with scientific fact and prophetic vision." The most important element of the stories was not the plot or the characters but the landscape in which the story took place, the technological setting. "Not only do these amazing tales make tremendously interesting reading—they are always instructive." wrote Gernsback. "They supply knowledge that we might not otherwise obtain—and they supply it in a very palpable form. For the best of these modern writers of scientifiction have the knack of imparting knowledge, and even inspiration, without once making us aware that we are being taught."

What was more plausible (or ridiculous), Gernsback asked his readers, Philip Francis Nowlan's story "Armageddon 2419A.D.," in which Anthony "Buck" Rogers was introduced into the twenty-fifth century, or W. Alexander's story "New Stomachs for Old," which suggested that one day organ transplants might be a commonplace surgical procedure? Gernsback delighted in pointing out that Jules Verne's *Twenty Thousand Leagues Under the Sea* had predicted the submarine "down to the last bolt" and that H. G. Wells had forseen the development of aerial warfare in his 1908 story, "The War in the Air." "New inventions pictured for us in the scientifiction of today are not at all impossible of realization tomorrow," even such terrifying ones as the apocalyptic bombs made from uranium in Wells' 1914 tale "The World Set Free."

By 1928 *Amazing Stories* had garnered a monthly circulation of well over 100,000 and a host of competitors had emerged: *Weird Tales, Miracle Science and Fantasy Stories,* and *Astounding Stories of Super-Science.* Initially an offshoot of the fantasy or adventure story, scientifiction had now become a distinct literary genre, albeit one which still made up only a small percentage of the pulp market and which was not taken very seriously by anyone outside its coterie of teenage admirers and a handful of adult enthusiasts.

For the most part scientifiction pulp writing paid abysmally. Between half a cent and one cent a word was the typical rate even for renowned authors such as Edgar Rice Burroughs and H. P. Lovecraft. A 6,000-word story could bring in as little as $30. Thus quantity tended to outstrip quality. Writers came from a variety of backgrounds. Jack Williamson, who published his first story in *Amazing Stories* in 1928 at the age of twenty, had grown up on a farm in New Mexico, living in the covered wagon that had transported him and his family there, while Roger Sherman Hoar, writing scientifiction under the name Ralph Milne Farley, was a Harvard-educated Massachusetts state senator. The common factor entwining reader and writer was a sense of wonder at the possibilities of science. "In my imagination" wrote the budding scientifiction author Jack Williamson, "science had always been magic made real, the promise of unlimited wisdom and power that even I might hope to learn and use." The young Parsons read *Amazing Stories* and the other scientifiction pulps religiously and would continue to do so throughout his life.

The most popular of the stories and articles to be found in *Amazing Stories* spoke of space flight. Although pulp writers had dreamed up huge guns, antigravity devices, and occult rays as possible ways to power their heroes across the universe, the rocket was slowly becoming identified in the minds of the pulps' readers as the tool most likely to actually get one aloft.

A Chinese bureaucrat named Wan Hu had attempted the first manned rocket flight in A.D. 1500. He built a chair balanced on two wheels, and to its base he attached forty-seven black powder rockets. Seating himself in the chair, he clasped a kite in each hand in order to sustain his flight once he was born aloft. With a glance at his assistants, Wan Hu signaled that the rockets beneath him be lit. As the fuses flamed, his chair became engulfed in smoke and fire. When the powder ignited, an explosion ripped through the air. The assistants craned their heads to the heavens, trying to see their master ascend to glory. He must have been successful, because when the smoke cleared, the chair, kites, and Wan Hu had disappeared, never to be seen again.

Few scientists tried to repeat Wan Hu's hands-on experiment, and although the rocket slowly began to appear in literature as a means of propulsion into space, it was never taken very seriously. In Cyrano de Bergerac's proto–science fiction tale, "L'Autre Monde" ("The Other World") (1657), the hero, Cyrano himself, straps a rack of rockets to the back of a winged machine he has made. He intends that as each rocket burns out, it will ignite the next, so renewing the boost and powering him towards the moon. The invention works and he is propelled into space, but his rocket machine falls apart around him. After falling for many days, he lands in the Garden of Eden, a fate denied the unfortunate Wan Hu. The rocket was thus seen as little more than a comic prop, a synonym for the absurd and impossible, the equivalent of an intergalactic banana skin.

However, scientifiction sought to establish a sense of plausibility around the idea of space travel and rocket ships. Indeed, many future rocket scientists would admit that scientifiction— or, as it was quickly becoming known, "science fiction"—was the catalyst for their interest in the science. "More than any other nation," writes the space historian, Frank H. Winter, "America traces its astronautical roots to a science fiction fatherhood." Parsons joined them in being romanced into the field of space travel. But how was a teenage boy to get started on Gernsback's space quest? Where could one enroll oneself in this new technological army?

There was little educational guidance for a boy wishing to begin experiments in interplanetary travel in the 1920s. Rockets may have been mentioned in some of the nonfiction articles that appeared in the pulps, but this by no means made them a subject for science. Indeed, the pages of the pulps were the only avenue open for such technical articles. There were no courses taught on rocketry, and there were few science magazines willing to devote pages to its discussion. Every American may well have sung about "the rocket's red glare" in "The Star-Spangled Banner," but in the century since that poem had been written, the rocket had fallen into disuse. It had become an object of ridicule.

This reaction was not against the shock of the new. On the contrary, rockets of one kind or another had existed for a millennium. In the eleventh century A.D., the author Tseng Kung-Liang wrote an account of his country's use of "fire arrows" in the war against the Mongols. To power these the Chinese used a combustible mixture of charcoal, saltpeter (naturally occurring potassium nitrate formed by the decomposition of animal and vegetable matter), and sulphur, previously used in fireworks: a combination we call black powder or gunpowder. A bamboo tube was filled with this powder and either pointed in the direction of the enemy or tied to the side of traditional feathered arrows. When the tube was ignited, the arrows could fly up to

1,000 feet. They proved devastating weapons, as much for their terror-inducing capabilities as for their effectiveness in slaughter. Indeed, by the thirteenth century, reports of rockets being used as weapons were emanating from Italy, Arabia, Germany, and England, and saltpeter had become as precious as gold.

Rockets remained very much a part of the blood and turmoil of earthbound activities for centuries. They were used to devastating effect against the British cavalry by the Indian Haider Ali's thousand-strong rocketry contingent in 1788. By simultaneously exploding rockets and firing rockets directly at the British troops, the Indian soldiers caused extreme panic and horrendous damage. The rockets were primitive: often little more than small iron casings stuffed with black powder and bound to metal swords. Others were simply sharpened bamboo tubes, sometimes six feet long, filled with powder and designed to bounce along the ground towards the enemy.

The British, chastened by this unseemly defeat, developed their own rockets through the pyrotechnical genius of Sir William Congreve. He created large ironclad fire bombs, 25,000 of which were used to burn Copenhagen to the ground during the Napoleonic War, and Britain used them again against the United States in the War of 1812. By 1815, all the great armies of the world included a rocket division. But by the middle of the nineteenth century these same divisions were being disbanded. Despite centuries of use as mayhem-making machines, rockets had barely improved since the Chinese fire arrows. No scientist had managed to completely understand their workings. They were unpredictable, often misfiring and almost impossible to aim with much precision. Guns and artillery, on the other hand, were improving remarkably in terms of both range and accuracy and it was not long before the rocket went the way of the crossbow and slipped into military obsolescence. At the Centennial celebrations held for the Peace of Aix-la-Chapelle in 1848, the British turned a large number of military rockets into fireworks to light up the River Thames. The consensus was that rockets were impractical for anything more than distress signals and fireworks, and the rocket was relegated to an amusement.

Nevertheless, the rocket's scientific basics were on display to Parsons just a short walk from Orange Grove. The easy availability of legal fireworks combined with the ingenuity of the local Pasadena boys had turned the Arroyo Seco into a blasting range. In one of the more popular games, boys placed powerful firecrackers under empty tin cans, lit the fuse, and then ran for cover. Peeking over rock barricades, they watched a barrage of explosions flinging the cans into the air. The boy whose can flew highest was the winner. It was a crude but effective lesson in Newton's third law of motion: "For every action, there is an equal and opposite reaction."

A rocket, at its most basic level, is little more than an example of this law. Chemicals are burned within an internal chamber, and the products of combustion—mostly hot gases—shoot out the only exit available to them, an opening in the back of the casing. The gases go one way and the rocket is compelled to fly in an opposite direction. With instructions most likely gained from one of the many "Build-Your-Own" kits advertised in the pages of the pulps, Parsons began to construct his first skyrockets.

In the sumptuous surroundings of his garden, Parsons, probably with the help of his indulgent grandfather, would scrape the explosive black powder out of the fireworks and cherry bombs he had collected. He would then tamp the explosive into a casing, probably fashioned out of wound paper or balsa wood. The harder the powder was compressed, Parsons would have noticed, the faster the rocket seemed to go. Attached tightly to the end of the tube was a clay nozzle, through which a paper fuse would protrude. He would tie a stick to the rocket's side and ram it into the ground, to act as a launching tower, then light the fuse. With a fizz and a roar the rocket would shoot into the sky, scorching the grass as it left, diminishing in size rapidly until a second later, its charge spent, gravity would drag the

empty casing back down to earth. It seems likely, judging from
his near addiction to rocket experiments in his later life, that
such launchings were a frequent sight and sound on Orange
Grove. The thrills of the explosion, the roaring launch, and the
sight of the zooming "bird" would remain essential ingredients
in Parsons' love of the science. One can only imagine the dis-
gruntlement of the members of the genteel Valley Hunt Club
next door as burning cardboard drifted down among them during their afternoon tea.

By 1926 Los Angeles was letting out one of the loudest hollers
of the Roaring Twenties. A visiting Aldous Huxley wrote of the
din of romping flappers, "barbarous music," and the "Gargan-
tuan profusion" to be found in the city's restaurants:

> How Rabelais would have adored it! For a week at any
> rate. After that I am afraid, he would have begun to miss
> the conversation and the learning, which serve in his Abbey
> of Thelema as the accompaniment and justification of
> pleasure. This Western pleasure, meaty and raw, untem-
> pered by any mental sauce—would even Rabelais's un-
> squeamish stomach have been strong enough to digest it? I
> doubt it.

Los Angeles' population tripled to 1,470,516 in the 1920s,
making it the fifth largest city in the nation. Oil had been discovered in Long Beach, twenty miles to the south of the city,
turning the Port of Los Angeles into the second busiest in the
country. However, the biggest success had been filmmaking.
With 90 percent of the world's films now made in the Greater
Los Angeles area and with over $247 million a year spent mak-
ing movies, it was the biggest industry in the city. Los Angeles
had become truly wealthy, and it was not shy about showing it.
Sumptuous resort hotels were built, containing giant dining
rooms and surrounded by lush golf courses that helped attract

some of the million-and-a-half tourists that visited the city each year. The city's bigger-is-better philosophy translated itself into cultural events, too. When a production of Shakespeare's *Julius Caesar* was performed in Beachwood Canyon, a cast of 3,000 actors was enrolled to act as the opposing armies in front of a crowd of some 40,000.

Sheltered from this tumult of hedonism and big business by Pasadena's lofty detachment, Parsons began attending Washington Junior High School at age twelve. A lack of school records up to this point suggests that his early schooling might have come from a tutor or governess, a form of education still fairly common among wealthy families of the time and all the more likely in Parsons' case since he seemed to suffer from a form of dyslexia. Throughout his life he would misspell words, and his handwriting in particular—the words usually printed in capitals rather than written in cursive—indicates a learning disorder. At the time dyslexia was not considered a legitimate complaint, and children who suffered from it were generally supposed to be backward or stupid. For anyone, let alone such an avid reader as Parsons, the variable grades that resulted from this learning disorder would have only fueled a dislike for establishment education.

His mother's pampering had made him a slightly plump child, and his solitary upbringing had led to his rejection by most other children. At Washington Junior High School, he was considered "effeminate" and was teased for having the politeness and manners of a rich "mummy's boy." The gaudy colors of his science fiction magazines also signaled him out for abuse; the playground was not the safest place to admit an interest in science. But he might have been spared the scorn of his peers had he not arrived so conspicuously on the first day of the school term in his grandfather's limousine. Immaculately turned out in a grey wool blazer, knitted brown tie, and leather shoes, he spoke with an affected English accent, no doubt picked up from the servants in his home. He stuck out like a sore thumb amidst the rough and tumble of school life. He became known as a "sissy" and was relentlessly taunted for his fancy clothes, while his long hair was grabbed and tugged by the school bullies. "Unfortunate experiences with other children," as he later referred to these incidents, led him to shy away from the crowd and devote himself to his books.

Edward Forman was almost two years older than Parsons and also suffered from dyslexia, but there the similarity ended. Forman was tall and good-looking, street-smart and affable, and he had a distinct streak of rebelliousness running through him. In Parsons' first year at the school, Forman was designated a monitor with the thankless task of summoning the pupils in from recess. One day a scuffle broke out on the playground. The students quickly formed a circle, shouting and yelling. Through the dust being kicked up, Forman could see someone getting a serious beating. He raced into the middle of the circle, pulled the bully off his victim, and with one well-aimed punch broke the assailant's nose. The fight was over. The crowd dispersed. Forman looked down on the ground where the bloodied and dirty figure of Jack Parsons, his books splayed out around him, was sprawled. He helped him up. So began the closest friendship of Jack Parsons' life.

Unlike Parsons, Forman was not from a wealthy home. Only a few years before his family had been farmers in Missouri. They had moved to Pasadena in search of a better life, but upon their arrival they had found no houses for rent and the grand resort hotels far too expensive. Along with his parents and three older brothers, Ed was forced to camp in the Arroyo Seco until a residence could be found. Forman's father found work as an electrical engineer and a house was eventually obtained. Even so, the family had to take in a boarder to help make ends meet. Forman "saw this rich boy whose grandfather gave him a twenty dollar bill every day," remembered Helen Parsons, Jack's first wife. At a time when the average wage was fifty-six cents an hour, the beneficence of Walter Whiteside hung like a "kick me" sign from Parsons' back. "Ed took Jack under his wing because he saw the help that he needed," said Helen Parsons. Later in his life Parsons would call his friendship with Forman "essential in developing" his "male center."

Parsons appreciated the advice and the protection Forman gave him, while for his part Forman not only enjoyed helping spend Jack's pocket money but also listening to the eloquent and well-read Parsons holding forth on any number of strange and mysterious topics. With the eagerness of a lonely boy, Parsons would have told Forman of the magical worlds of Parsifal and Sir Gawain, of eastern religions and outer space. Forman was no stranger to this last subject. He was already an avid reader of Edgar Rice Burroughs' series of books on Mars, in which the hero, John Carter, falls asleep in a cave in the Arizona desert and through an unexplained mystical process finds himself waking on the red planet. It was not long before the two boys were spending their spare time reading and earnestly discussing science fiction together.

It was at this time in his life that Parsons would later claim to have had his first mystical experience: He attempted to invoke the devil in his bedroom. He would describe the experience later as his "magical fiasco," which put him off further occult study until he was older, but he also intimated that he had succeeded and scared himself witless. If he had mentioned the story of the devil to Forman, one can only imagine how it might have intrigued the older boy. "I think Ed just worshipped Jack," remembered Jeanne Ottinger, Forman's stepdaughter. "Ed was bright, very bright, but he just didn't have the formal education that Jack did. He learned a lot from Jack and I think Jack and he were exactly the same kind of adventurous person."

Above all it was Parsons' interest in rocketry that captivated Forman, and between Parsons' pocket money and Forman's engineer father, the two would have plenty of materials to work with. "It was our desire and intent," remembered Ed Forman, "to develop the ability to rocket to the moon." The pair adopted the phrase *Ad Astra per Aspera*—through rough ways to the stars—as their motto. They swiftly became inseparable as they drove each other on to create more complex and explosive sky-rockets, the balsa wood tubes growing larger, more aerodynamic, sprouting fins and nose cones just like the rockets they had seen pictured in the pulps. They seem to have made efforts to replace firework powder with an explosive even stronger, for by the time the two moved to Pasadena's John Muir High School in 1929, they had gained a reputation for mischief. "They were a couple of powder monkeys," remembered Marjorie Zisch, a fellow pupil at John Muir. "They would go out into the desert and make rockets and do all sorts of explosive stuff." With the exception of Forman, Parsons still had few friends, and he made little effort to fit in. While John Muir prided itself on its football, basketball, baseball, and track teams (the great baseball player, Jackie Robinson, was only a few years behind Parsons at the school), he preferred fencing and archery—solitary sports surrounded by an air of old-world romance.

At some point during his teenage years, however, his increasing fondness for explosions and his poor school grades began to worry his mother. Whether Ruth Parsons thought that her boy needed to be toughened up, or whether she hoped to tame his increasingly volatile enthusiasms, she decided that Brown Military Academy for Boys, 130 miles to the south in San Diego, would be the best place for him. She could not have been more wrong. For a boy who naturally shied away from groups and regimentation, the forty-acre academy known as "the West Point of the West" was unfortunate in every respect. He was dismayed to find the school rife with bullies, and his protector Forman was far away in Pasadena. If the academy taught him anything, it was that the practical application of explosives could cause a rapid reaction. "He blew up the toilets in the whole goddamn place," remembered Jeanne Forman, Ed Forman's future wife, and was promptly sent home again. Helen Parsons recalled, "They were trying to make a man out of him and they got a donkey."

Back in Pasadena, he renewed both his friendship with Forman and his rocketry experiments. At school the other pupils did not make fun of him anymore. He had gained the confidence that only being expelled from a military academy can give one. What's more, he was rapidly becoming a good-looking boy, with his long black hair greased straight back and "glowing, piercing" eyes.

Parsons' and Forman's minds may have been fizzing with the idea of rockets, but the world itself was quite indifferent to thoughts of travel to the moon. Instead, the airplane and its supporting science of aeronautics ruled both the skies and dreams. In the 1920s the burgeoning new business of aviation had chosen Southern California as its center. Attracted by the promise of 350 clear flying days a year and the ability to park planes outside, pioneer aviators such as Glenn Martin, Donald W. Douglas, John Northrup, and Allan and Malcolm Loughead (later Lockheed) set up shop around Los Angeles. They were young men in their early thirties who happily worked together building planes in disused movie lots with the backing of wealthy aviation enthusiasts. They initially sold their aircraft to the navy or to the postal service, which needed mail carriers. However, by the late 1920s the development of passenger lines offering flights from Los Angeles to San Diego, Seattle, San Francisco, and Salt Lake City had increased demand rapidly. More and more aircraft were being built, and Los Angeles had become the undisputed capital of the aviation industry in America.

Adding a further thrill to this budding industry was the glamour that surrounded flight. When in 1927 Charles Lindbergh succeeded in the first solo, nonstop airplane flight across the Atlantic, he made aviation the adventure of the day. In Los Angeles the appeal of the airplane had been amply demonstrated by the millionaire playboy, Howard Hughes, who had begun filming his First World War aerial masterpiece *Hell's Angels* in the skies above. Having amassed the largest private air force in the world, he was now filming aerial dogfights along the coast, wowing the populace and firmly establishing the airplane as the preeminent awe-inspiring technology of the day.

The government and the academic world followed suit. As universities built science departments, the propeller-driven airplane was seen as the harbinger of the new technological age, receiving ample support. However, there were no compelling economic, military, or scientific reasons to study the rocket, which at the time was used solely to propel lifelines aboard ships and occasionally whaling harpoons. Thus, anyone who saw possibilities in the use of rockets had to work alone, driven solely by personal passion.

Unsurprisingly, those that did study these strange and impractical engines were not cut from the usual scientific cloth. In 1881, a twenty-seven-year-old Russian explosives technician named Nikolai Kibalchich sketched and described a flight vehicle propelled by a solid-fuel rocket. The rocketeer envisioned a moveable rocket engine attached to a platform, which would allow the craft to be steered by adjusting the direction of thrust of the engine. "I think that in practice, such a task is achievable . . . and can be accomplished with modern technology," he wrote. His farsightedness was made all the more remarkable by the fact that he was being held prisoner in the Petrapavloskaya Fortress in St. Petersburg, awaiting execution for creating the bombs that had been used in the assassination of Emperor Alexander II. "I believe in the reality of my idea," Kibalchich wrote, "and this belief supports me in my terrible situation."

It was not until 1903, however, that the greatest step towards treating rocketry as a science was taken. Another Russian, Konstantin Tsiolkovsky, an impoverished, deaf schoolteacher, inspired by the stories of Jules Verne, published his classic treatise *The Probing of Space by Means of Jet Devices*. Developed through a series of experiments made in his home laboratory, his theoretical groundwork dealt for the first time with such complex problems as escape velocities from the earth's gravitational field and the relationship between the mass of the rocket and its propellant. Although this impressed a small clique of physicists in St. Petersburg, his work would remain little known outside his native country until the 1930s.

America had its own rocketry pioneer, Robert Goddard. Born in 1882 in Worcester, Massachusetts, Goddard was the son of a bookkeeper and a disinherited merchant's daughter. He was nurtured in his youth on the stories of H. G. Wells, in particular his novel of a Martian invasion of earth, *The War of the Worlds*. The tale of Martians traveling over 140,000,000 miles through outer space impressed him immensely, as did Wells' "compelling realism" in the telling of the story. Then, on October 19, 1899, while climbing a cherry tree at his Massachusetts home, he experienced a scientific awakening on a par with a religious epiphany: "As I looked toward the fields at the east, I imagined how wonderful it would be to make some device which had even the *possibility* of ascending to Mars, and how it would look on a small scale, if sent up from the meadow at my feet." In front of him seemed to materialize a mechanical device, as solid as the tree he sat in, that whirled round and round until it began to lift, twirling and spinning above the city of Worcester and out into space. "I was a different boy, when I descended the tree from when I ascended," he wrote; "existence at last seemed very purposive." Since that tumultuous day he had gone on to devote the rest of his life to what he saw as "the most fascinating problem in existence," rockets and space travel, or, as he grew to call it, "high altitude research."

Operating largely by himself while teaching physics at Clark University, he experimented extensively with his own black powder devices, before publishing in 1919 the founding text of modern rocketry, *A Method of Reaching Extreme Altitudes*.

Among much dry description of his earthbound experiments, he hypothesized that a rocket could be used to attain sufficient velocity to escape from the earth's atmosphere. Its success could be demonstrated by crashing the rocket onto the moon with a payload of flash powder which would signal its arrival to watching astronomers.

Goddard wrote his text to gain funding for more experiments from the Smithsonian Institution, and he included the moon rocket hypothesis purely as an illustration of his more abstruse calculations. However, when it fell into the hands of the newspapers, it created a sensation. AIM TO REACH MOON WITH NEW ROCKET, read a headline from the *New York Times*. MODERN JULES VERNE INVENTS ROCKET TO REACH MOON, read the *Boston American*. For a brief moment the rocket took over the nation's fancy. Goddard began to be called the "moon-rocket man." Novelty songs such as "Oh, They're Going to Shoot a Rocket to the Moon, Love!" were written about him. He was mocked and attacked in science journals for his idea. The *New York Times,* in a woefully misinformed editorial, chastised the professor for not knowing the "relation of action to reaction, and of the need to have something better than a vacuum against which to react ... Of course he only seems to lack the knowledge ladled out daily in high schools." (The writer of this editorial failed to understand the critical third law of motion, the one even the twelve-year-old Parsons had grasped: "For every action, there is an equal and opposite reaction." The fact that the reaction takes place in a vacuum is irrelevant.)

For the shy and retiring Goddard, this was too much. He became more secretive and hostile to enquiries. His rockery work progressed, but he would not share his hard-won secrets even with his admirers. In 1930, at the age of forty-eight, he suffered the added humiliation of being forced to move his home from Massachusetts, to Roswell, New Mexico, after the sound of one of his rocket experiments was reported to the police as a plane crash. Like a prophet he retreated into the wilderness. His life became a cautionary tale of the scorn the study of rocketry could command.

Along with Hermann Oberth of Romania, who wrote *The Rocket into Interplanetary Space* in 1923, Tsiolkovsky and Goddard gave mathematical form to space flight for the very first time, though none of them knew of each other's work. Tsiolkovsky dealt with the fundamental laws of motion in space, Goddard made calculations on the amount of solid propellant needed to power a rocket, while Oberth suggested liquid fuels as a means of propulsion and considered the hitherto unexamined problems of space suits, space walks, and the minutiae of embarking on long distance interplanetary journeys. They would be the forefathers of the field that would become known as *astronautics*—the science and technology of space flight—a term, needless to say, that was invented by a science fiction writer in 1927.

The work of these three pioneers was the theoretical catalyst for the rocket societies that had begun to flourish worldwide in the late 1920s and 1930s, generating tiny pockets of feverish enthusiasts enraptured by this strange science in Argentina, Germany, Austria, the United Kingdom, Russia, France, and Japan. While these societies were considered little more than a joke by the popular media and beneath contempt by the academic community, they paid close attention to each other and to what the other members of their far-flung community were achieving.

On the evening of April 4, 1930, one such group of eleven space-minded men and one woman met for the first time in a small brownstone building in New York City. They ambitiously called themselves the American Interplanetary Society (AIS) and unashamedly stated their ambitions to promote "interest in and experimentation toward interplanetary expeditions and travel." They would become one of the few guiding lights in these dark days of rocketry research, making contact with other international rocket groups and expanding into a society of great renown by the 1960s. Their beginnings, however, were modest at best. The group had come together largely because nine of them were science fiction writers, editors, or rewrite men for Hugo Gernsback's latest science fiction magazine, *Science Wonder Stories*. They included the bearish Edward G. Pendray, a *New York Herald Tribune* reporter who wrote for Gernsback under a pseudonym; his wife Leatrice, a nationally syndicated woman's page columnist; Warren Fitzgerald, head of The Scienceers, a multiracial science fiction fan club based in Harlem; and Dr. William Lemkin, a chemist and the only Ph.D. among the group. Gernsback himself joined the society but did not attend meetings, preferring to skim through the minutes in order to garner story ideas for the next edition of his magazine.

The society had an infectious optimism and naïveté. "It was our expectation," remembered Pendray, "that engineers and scientists would spring to our service if we but called their attention to the possibilities of rockets in an appropriate manner." They could not have been more wrong. They sought members through advertisements in the pulps and through their own mimeographed "Bulletin of the American Interplanetary Society," and attracted young enthusiasts like Parsons and Forman, who joined immediately. But they lacked funds and scientific literature, and no matter how hard they tried, they could not interest the wider world in their quest. Most dishearteningly of all, Robert Goddard flatly refused to help them. Goddard had recently gained financial backing for his clandestine rocket from the aeronautically-obsessed Guggenheim family, and such patronage he did not want to be disturbed. So it was that the one man who had more insight into the possibilities of space travel than anyone else would not share details of his work with those who most wanted to make it happen.

In Germany, the nascent rocketeers were having more luck. The Society for Space Ship Travel, or Verein für Raumschiffahrt (VfR), was founded in 1927, and was slightly more professional than the AIS, if equally idealistic. The society had been started by an odd assortment of engineers and clergymen interested in space flight, but their early slogans such as, "Help create the spaceship!" quickly drew a membership, particularly among the underemployed engineering community still reeling from the devastating economic aftershocks of the First World War. Most significantly of all, the VfR enticed Professor Hermann Oberth to become its president.

If one man could be said to have rivaled Goddard for his rocketry skills it was Oberth. Fascinated equally by reincarnation and rocketry, in 1923 he created a stir with the publication of his doctoral thesis *Die Rakete zu den Planetenrämen (The Rocket into Interplanetary Space),* which demonstrated through elaborate mathematical proofs that rockets could be built to transport man far beyond the reach of Earth's gravitational pull. By the time the AIS formed, the VfR already had a *Raketenflugplatz* (a rocket test site) in an abandoned army garrison, and their experiments were well underway. With such future greats as the seventeen-year-old Wernher von Braun (who would eventually develop rockets for America's manned lunar program) among their ranks, the VfR was the preeminent rocket society of the age.

Nevertheless, the enthusiasms of its young members, many of whom were still in their teens, often overpowered good science. Like Parsons and Forman, most of them wanted to see their rockets fly. Launchings would take place before designs had been properly calculated, often resulting in chaos. In a 1931 letter to the AIS's Pendray, Willy Ley, one of the club's founders and later one of America's best-known spokesmen for space exploration, told of how the VfR had "destructed a house of police" with a rocket that had gone astray and landed on the roof of the local police station. It prompted a temporary ban on all experiments.

Curious to learn more about the German experiments that he read of in the AIS's bulletin, Parsons wrote directly to Wernher von Braun, telling him of his own primitive experiments and asking for more information on building rockets. Ed Forman's third wife Jeanne remembered how both Parsons and Forman had called Braun by telephone, presumably at Parsons' grandfather's expense: "Both of the boys had talked to von Braun many times by telephone, way before he ever came to the U.S. . . . They were crazy about von Braun and he was crazy about them, because they were out horsing around with the same stuff."

The members of the rocket societies on both sides of the Atlantic were intent on extracting information about experiments and advances through letters and telephone calls. With such a small number of people seriously interested in the science, every bit of knowledge needed to be shared. Braun, who was only two years older than Parsons and a science fiction fan as well, did his best to reply to Parson's entreaties, and he also asked questions of his own. Parsons, however, increasingly found himself frustrated by the data he received back; indeed, Forman suspected that he and Parsons were being asked to reveal a little too much of their own experiments. They severed their correspondence. Even at this nascent stage, it was possible to feel proprietorial over their work.

The events of 1929 gave the world even more reason to disdain wistful thoughts of travel to the moon. How could one think of rockets when the *Graf Zeppelin*, a 776-foot-long passenger dirigible, stopped in the city on the final leg of its journey to circumnavigate the globe in twenty-one days? One hundred fifty thousand people flocked to watch it land, and a giant banquet was held in its honor. Attendees included a host of film celebrities such as Douglas Fairbanks and Mary Pickford, as well as the state governor, the mayor and the media mogul William Randolph Hearst. It was the "single most glittering event to date" in Los Angeles' history. The enthusiasm and optimism of the 1920s were at their peak.

In Pasadena, Walter Whiteside was planning his own starring moment. He wished to sell his beautiful house on Orange Grove complete with all its furnishings in order to build a new one, on the precipitous western side of the Arroyo Seco. Its construction funded by the success of his real estate investments, this new house would be Walter Whiteside's Xanadu. It would look back the half mile to the meandering Colorado Street Bridge that spanned the Arroyo and led to the city's steeples and trees beyond. Parsons' grandfather would lord it over not just the Valley Hunt Club, but the entire city of Pasadena.

The family traditionally vacationed on Santa Catalina island, twenty miles off the coast of Long Beach, which William Wrigley, a fellow Pasadena resident, had transformed into a resort. In the summer of 1929, however, the family embarked en masse for a long trip around Europe while construction proceeded on the new home. It would be the only time that Parsons traveled abroad in his life. While his grandparents searched for ornaments to fill their new mansion and Ruth Parsons availed herself of the latest Paris fashions that made her "the best dressed woman" in Pasadena, the adolescent Parsons would stand for hours at the backstage doors of the theaters in Montmartre and hope to catch himself a showgirl. His growing charisma and considerable allowance compensated for his slight grasp of foreign languages. The family returned to the sight of the house at 285 North San Rafael Avenue, sitting triumphantly on the gorge's edge. They quickly stocked it full of new European treasures.

But even the luxuriant verdure of Pasadena could not entirely protect its wealthier residents from the Wall Street crash that occurred in the autumn of 1929. Southern California suffered the highest bankruptcy rate in the country, and although Walter Whiteside did his best to keep up his standard of living, within two years his fortunes had ebbed away. The limousine which dropped Jack off at school each day disappeared, and Ruth Parsons began to work as a shop assistant back in hated Los Angeles. Even the view from the house had been spoiled. In the four years after the crash, seventy-nine despairing investors plunged to their death from the graceful curves of the Colorado Street Bridge. It swiftly became known as Suicide Bridge.

The crash also triggered a decline in Walter's health. Having attained his Xanadu, he ruled it for less than two years before death claimed him in July 1931. It was a great blow to the young Parsons. His grandfather had been the closest thing he had to a father. Writing about this influence later, he explained that his grandfather had been essential in preventing "too complete an identification" with his mother. Now he was to be the sole man in the house.

Whether because his writing disability required special attention or because of disciplinary problems similar to those that had caused his removal from military school, Parsons left John Muir High School in 1931, as did Ed Forman. Forman dropped out of school completely, beginning a round of odd jobs as a carpenter, chauffeur, and postman. But the last of the Whiteside fortune went towards sending Parsons to the University School, a private all-boys establishment in Pasadena. While the school occupied a large, white colonial mansion, it was perennially bankrupt. It had enrolled only thirty pupils when Parsons arrived and these consisted mostly of the sons of wealthy families who had been expelled from other schools. An old wino was allowed to live on the school grounds in a shack where he obsessively took apart and rebuilt an old motor. The school's headmaster and proprietor was Russell Richardson, a passionate liberal who regularly attended meetings of the American Federation of Labor as well as the visiting British philosopher Bertrand Russell's lectures on such subjects as "Is Monogamy Doomed?" Richardson was a keen proselytizer for new educational methods. Speaking to the *Los Angeles Times* in 1929, Richardson praised the rebellious mind-sets of his pupils: "These young people have an immense energy that will do great things if properly directed...Their attitude of not accepting everything in the classrooms with blind faith, their questioning of all values, means a closer contact with the currents of life than could possibly have been true of a previous day." For Richardson, "the personality of the teacher counts for more than the methods he uses" and "the best kind of development is self-development." Richardson's teaching was not conventional, but it suited the young Parsons perfectly. At the University School he flourished. He won an award for literary excellence and became one of the editors of the school newspaper, *El Universitario*. His greatest interest was in chemistry, particularly when he saw how knowledge of chemicals could help him concoct more powerful fuel for his rockets. For this subject he was in good hands: Many of the school's teachers came from the adjacent California Institute of Technology.

His time there was marred by only one unfortunate incident. The school's assistant headmaster and chief disciplinarian was a former military man, Captain John Miles. His ability to keep the more unruly pupils on the straight and narrow through his army braggadocio was much relied upon by Headmaster Richardson, and many of the pupils, including Parsons, had a great deal respect for him. He was also vain and boorish, with a volatile temper. When Richardson attempted to replace him with another, less expensive army man, Miles was furious, as were many of the students. Parsons' position as the school's literary editor made him a natural choice to be the de facto spokesman for the complaints of the pupils. In his strangely clipped writing, he wrote a petition for the other pupils to sign:

> We the following students of the University School, wish to express our appreciation to Captain Miles for his help and cooperation during the past two years. We extend to him the sincere wish that he always receive the same square deal that we have received from him.

Underneath the seventeen signatures, representing the senior graduating class of the University School, he wrote, "Drawn by J. Parsons at the request of several students." Parsons presented the letter to the headmaster as a direct challenge to his authority. The letter was not one to be taken lightly. With so few pupils, the departure of even one of the "young gangsters" (as Mrs. Richardson described them) would have had devastating consequences on the school's finances. But the rebellion appeared in another light a few days later when a penitent Parsons confessed to the Richardsons that Captain Miles had "forced" him to write the petition. Instead of being the head of the rebellion, he had merely been a pawn in someone else's power play. In her journal, Mrs. Richardson writes of certain of the signatories as "problem boys ... glad to gang-up. Others easily led by these toughies." Parsons avoided the "problem boys" tag, but his tendency to be easily led by unscrupulous others would remain an indelible character flaw.

By 1932 the depression was at its worst. Twelve million people were unemployed in a country of 123 million, and the middle class were in danger of being completely wiped out. No new mansions had been built on Orange Grove since 1929, and now the giant homesteads were groaning under the weight of their domestic staff. Maids, butlers, cooks, washerwomen, chauffeurs, and Japanese gardeners were all reliant on the dwindling fortunes of the mansion owners. As the money disappeared, a slow trickle of unemployed workers began to flow out from Orange Grove.

Parsons, eighteen years old and the sole man in a family of two widows, had one more year of school to go, but the family's finances spurred him into action and out of the dreamworld he had been occupying for much of his childhood. "The loss of family fortune developed your sense of self-reliance at a critical period," he wrote about himself in later life. "The contact with reality at this time was essential." He found the perfect part-time job at the office of the Hercules Powder Company in Los Angeles. Hercules offered a treasure trove of explosives information and Parsons' curiosity was boundless. The company made ammonium nitrate for coal stripping, nitroglycerine for ditch blasting, gelatine for shaft sinking, and ammonia dynamite for road building; there had been a strong powder industry in California ever since its mining heyday, and every type of explosive, from firing caps to powder kegs, was on display and easily accessible.

Learning explosives lore from the other workers, Parsons soon discovered such essentials as the difference between a high explosive and a low explosive. A high explosive such as nitroglycerine (the base constituent of dynamite, made by treating a natural by-product of the soap-making process, glycerine, with sulphuric and nitric acids) decomposes into gases in a few millionths of a second, about a thousand times faster than a low explosive such as black powder or gunpowder (traditionally made from a mixture of potassium nitrate, sulphur, and charcoal). Because of their rapid and violent detonation, high explosives are better suited to demolition work, while low explosives such as black powder are better used as a propellant, pushing projectiles out of gun barrels. Parsons was taught that making a good propellant is much harder than making a good high explosive, just as it is more difficult to sing softly or dance slowly, and he learned what ratio of ingredients, powder grain size, and density of packing would obtain the most powerful and the most reliable black powders. Parsons' discoveries were extremely important if he was to work out the best fuel for his rockets, and indeed it seems he often appropriated some of the company's explosives for his own private use. As he worked at the company on weekends and during the school holidays, his hobby slowly became a profession.

When he was not working at Hercules or going to class, Parsons would continue his experiments with Forman. Their working partnership was flourishing. Parsons would sketch drawings of the rocket and prepare the fuel, while Forman, his skills as a mechanic growing by the day, would build the rocket's outer shell. "Ed created Jack's thoughts," recounted Helen Parsons; "he manifested what Jack would say." Their aims were as lofty as any of the rocket societies: they wanted to create a rocket that could get to the moon. This ambition was often replaced by their enthusiasm to see the rocket fly regardless of what progress might be gained by it. The Parsons' backyard rapidly became as scarred and pockmarked as the lunar surface itself, and the two young men frequently drove out to the desert or wandered down into the Arroyo Seco to execute their tests.

Parsons graduated from the University School in the summer of 1933. In his yearbook he listed his theme song as "You Rascal, You." In the immediate months after his graduation, he helped his mother and grandmother move to a modest new house on a street that ran parallel to Orange Grove. The portly, shy young adolescent who had left Orange Grove four years earlier had been transformed. Standing six foot one, he was solidly built. His brown hair had begun to curl and he now kept it cut short. He was handsome, easygoing, and increasingly self-reliant. He still read classical literature voluminously and had also begun writing his own poetry. He would often declaim his favorites, in stentorian tones, to anyone who would listen. Robert Rypinski, an automobile dealer in Pasadena, struck up an instant friendship with Parsons when he came in to buy a used car.

> I wouldn't call his a forceful personality, you just felt that there was something…I've read about people who burn with a hard gem-like flame. Jack was that kind of a person to me, except it was…a warm gem-like flame.

Parsons enrolled in Pasadena Junior College in the autumn of 1933, hoping to earn an associate's degree in chemistry and physics, but he was forced to drop out after just one term. With the Whiteside family fortune in a perilous state, he could not afford to continue. It was a time when the majority of young people went to work after high school, or left high school to work full-time as Forman had. But a university education was essential if Parsons was ever to progress in the field of chemistry and explosives. Fortunately, however, he had impressed the management of the Hercules Powder Company in Los Angeles and, with the help of an old family friend, he got a job at Hercules' main explosives manufacturing plant in Pinole, on the eastern shore of San Francisco Bay, eight hours to the north of Pasadena. It would be a dangerous, exhausting job, but it would pay Parsons the significant sum of $100 a month: enough, if he was careful, to pay for a university education. What's more, he could see if the nearby Stanford University and University of California at Berkeley might be interested in his skill.

The Hercules plant was a community cut off from the outside world, and the buildings were constructed in a series of gullies and ravines. It was a desolate place. "At night the landscape around looks like a scene in Hell," he wrote in one of his many letters home, "muddy molten slag flowing downward—fan-like flames reddening the sky." It had been the largest TNT production plant in America during the First World War, when it produced over seven million pounds a month, but such work did not come without its risks. On average one worker died each year, usually in the devastating blasts that occasionally ripped through the nitroglycerine and dynamite buildings.

Regular work was a new experience for the eighteen-year-old. "The job started this morning, hard work—pushing four-ton dynamite cart, over acres and acres," he wrote in a letter home. But he was happy. "I'm glad it's hard—the iron must come out." Soon he was boasting of being able to "throw 100-pound dynamite boxes around all day." There were, however, unpleasant side effects to the work. When absorbed through the skin, nitroglycerine dilates the blood vessels, causing severe headaches. It was not long before Parsons was suffering from one. "I've got one of those damned nitro-glycerine headaches," he wrote on his second day at the plant; "they claim [it] will wear off in about a week."

Parsons took to the treacherous work environment with something approaching nonchalance. "Somebody fell into the jelly mixer today but due to the grace of God and his extreme

rotundity they fished him out before he got stamped into a car-
tridge and sold for a stick of dynamite." He seemed unflappable
even when averting a near catastrophe, as he did when preventing an unattended vat of high explosive from overheating:
"Wandering into clouds of smoke I shut off the machine and sat
down to an uninterrupted hour of rest until the crowd came
mournfully back to collect the pieces." Although Parsons may
have seen this danger as amusing, his mother soon became frantic with worry over her son's work, leading him to try and pla-
cate her as best he could with a mixture of humor and
self-assurance. "No accidents ever happen here except when
someone gets powder under his nails and lights a match," he
wrote. "But that usually turns out to be a good thing—excavat-
ing as it does a two year vegetable garden."

It was obvious that Parsons was a different type of worker
to the others who worked at the plant, "tough-hardboiled-but
good guys." Soon his knowledge of chemistry, his good manners and educated demeanor saw him playing contract bridge
with the chief chemist and the superintendent of the plant. Parsons claimed his preferment was the result of "hard work and
the old personality apple polisher."

His future, though, was uncertain. It had been suggested he
could be a superintendent of the plant, but his ambitions were
higher. "I don't want to be superintendent—I'd rather do the
work I've always aimed for—research and college. How do I get
there from here?" He became disillusioned with working sixteen
hours a day. His car proved to be his only means of escape, and
he often drove up into the hills overlooking San Francisco Bay,
where he watched the sunsets and the stars and longed for home.

Parsons was delighted when his promise was recognized by
the nearby Stanford University and he was offered a place to
study chemistry, but his joy was short-lived. He found the expense "higher than I first estimated." The money he had saved
from his Hercules job could not possibly pay for him to study
there for any amount of time, especially not with his mother and

grandmother to think about. He decided to return to Pasadena, where he could save money by living at home. "Farewell to Hercules," he wrote, "but the seven labors aren't completed yet."

One of the benefits Parsons took from Hercules was a near-encyclopedic knowledge of chemicals, of explosives in particular. He had an affinity with chemicals—what his friend Robert Rypinski called "that genius way of feeling . . . the substance of what went on in a chemical reaction." He now read chemistry books as keenly as he read the increasing number of science fiction magazines that were appearing on the market, and he was developing what a scientific colleague would later term a "global" grasp of chemistry theory. While Parsons had been away, Forman had worked as an apprentice to the machinists at Hercules in Los Angeles, creating shells and repairing gun machinery. The two now began fashioning their rocket engines from metal as opposed to the wood and cardboard rockets of before, correcting flaws in the rocket and the fuel through trial and error.

Nevertheless, they seemed to have hit a brick wall. To evaluate the different strengths and weaknesses of Parsons' fuels, they needed to measure the rocket's thrust, but they hadn't the equipment or the mathematical wherewithal to do this. They were amateurs, just "technicians in the powder business." Parsons confided to his friend Rypinski that they had "gotten out of their depth" and they needed someone "who could do some calculating for them."

Parsons and Forman were realizing that the rocket was a much more complex machine than either had first assumed. In its simplest form, the rocket resembles the internal combustion engine—that is, an engine powered by the explosion of gases in a cylinder. The combustion unit in a rocket, called the engine, or as Parsons would call it, the motor, is usually a metal cylinder distinct from the shell that surrounds it. It is supplied with a fuel, for example charcoal or gasoline, and an oxidizing agent, like potassium nitrate or liquid oxygen, which gives the fuel oxygen with which to burn. (Unlike jet engines, which burn oxygen from the atmosphere, the rocket carries its own oxygen within it, making it capable of traveling beyond the atmosphere.) When the fuel is ignited, large quantities of very hot gas are produced, expanding until they rush out of the back of the rocket, causing thrust. Anything that slows the exhaust slows the rocket, so a nozzle is attached to the rocket engine to direct the exhaust outward as efficiently as possible. Unlike the internal combustion engine, the heated gas acts upon no moving parts. As seen before, this property of the heated gas is in fact the purest expression of Newton's third law of motion. In the case of a rocket, the action is the backward-streaming flow of gas and the reaction is the forward motion of the rocket. Thus, the movement of a rocket does not depend on anything outside the motor: A rocket is not propelled forward because its exhaust pushes against air. In fact, a rocket actually works *better* in a vacuum as there is no air to impede the exhaust.

While a rocket's reaction principle is simple, its application is not. It is not just a crucible into which elements are thrown and set alight. Its success depends on understanding variations in the nature of the fuel—the speed and power at which it burns—the ability of the rocket motor to withstand the pressures inside it, and the overall design of the motor. Making a rocket that will fly where one wants it to at the speed one chooses demands a firm grasp of mathematics, chemistry, and engineering.

What Forman and Parsons wanted above all was the chance to build a rocket resembling the most recent work of the American Interplanetary Society and the VfR. The AIS by now numbered over one hundred members. Some were engineers who joined out of their enthusiasm for a challenge; other members signed up out of a profound despair brought on by the depression. "It was a lousy planet," remembered one member; "the rocket ship was the only way to get off it."

The Germans had managed to launch a rocket 640 meters into the air back in 1931 and shared their designs with the AIS.

The AIS used them to craft the society's first success. Parsons and Forman had read with envy of the construction of the AIS's experimental rocket No. 2. Fashioned out of a cocktail shaker, discarded aluminium coffee pots, valves liberated from an old cooking range, and fins of balsa wood, it stood an imposing seven feet tall, with two thin aluminium tanks containing the fuel (gasoline) and the oxidizer (liquid oxygen) flanking the rocket motor. The gasoline and oxygen were to flow continuously from these tanks into the rocket motor and there be ignited, thrusting the rocket into the sky. On the beach at Marine Park, Staten Island, on May 14, 1933, the rocket had been launched—but not without some drama. The "Bulletin" explained that the rocket's ignition system had failed to spark. One of the AIS members leapt over the protective sandbags and lit the rocket by hand with a gasoline torch. Before he could get back to safety, the rocket ignited, enveloping him in flame, then shooting some seventy-six meters into the air before its oxygen tank burst. To Parsons and Forman this sounded both terribly exciting and the logical next step.

In their report on the launch, the AIS had stated their belief that the future of rocketry lay in liquid propellants. The reason for this was simple: If you wanted to stop a liquid-fuel rocket motor from firing, all you had to do was stop the pumps that sent the fuel in. To control a liquid-fuel rocket's speed, you just had to adjust the rate at which the fuel and the oxidizer flowed into the motor. Solid-fuel rockets, by contrast, could not be stopped once started, nor their speed controlled once fired. Liquid fuels were also more powerful than solid fuels. But where were Parsons and Forman to get their hands on liquid oxygen, fuel injectors and, more importantly, the mathematical means that could help them calculate the forces being activated within the rocket motor? These could not be stolen from the powder companies. It seemed that they could not take the next step on their own. Fortunately, Pasadena, in its natural abundance, could provide just the help they needed.

3

•••••••••••

ERUDITION

A little learning is a dangerous thing.
—ALEXANDER POPE, *An Essay on Criticism*

In the science fiction writer Robert Heinlein's 1941 story "Universe," he describes a spaceship, big enough to hold and provide for whole families, that has gone astray. Generations have passed since the spacecraft's launch and the benighted descendants of the original crew have come to believe that the ship is the universe and that any notions of a voyage are to be understood in a purely religious sense. Those called "scientists" are, in fact, priests. When the open-minded hero of the story discovers the ship's main control room and sees the stars outside, he is filled with a quasi-religious ecstasy. Spiritual and technological revelation become one and, in the manner of Paul's conversion on the road to Damascus, the hero undergoes a glorious rapture.

Heinlein's story is symptomatic of the years in which it was written; science never held such a grip on the world's imagination as it did in the first half of the twentieth century. This was largely due to the work of one man: Albert Einstein. Einstein became the world's most celebrated scientist after his General Theory of Relativity, which predicted that light rays are bent by the gravitational field of large masses, was substantiated by the British astronomer Arthur Eddington's observations of solar eclipses in 1919. EINSTEIN THEORY TRIUMPHS read the *Times* of London; SPACE CAUGHT BENDING cheered the *New York Times;* and Einstein was thrust into the world spotlight as the harbinger and foremost proponent of the new scientific age.

It was an age in which a new universe was being illuminated. The German and Austrian physicists Werner Heisenberg and Erwin Schrödinger had, in 1926, laid the theoretical foundations of the new quantum mechanics; though their ideas violated the classical notion of causality—the belief that one object cannot influence another one without involving intermediate agents joining the two objects in space—they successfully predicted the behavior of atomic particles. In medicine the British chemist Alexander Fleming had recently created the first antibiotic when he discovered that penicillin could kill bacteria. Americans in particular were at the forefront of scientific discovery. In the famed "Great Debate" of 1920, the astronomer Harlow Shapley convinced his colleagues that the galaxy was at least ten times larger than had previously been believed. The physicist Karl Jansky discovered radio waves emanating from the Milky Way, thus opening the door for radio astronomy—the study of the universe through radio waves—while the electrical engineer Vannevar Bush and two MIT associates created in the "differential analyzer" the first modern analog computer.

In Pasadena, at the California Institute of Technology, the reverence for the scientist seemed particularly strong. White terracotta classrooms and subterranean laboratories seemed to hold the promise of technological revelation, and the Institute's arcaded lawns and Romanesque revival cloisters emanated "a near-religious dedication to science and to the priesthood of scientists."

A strange mix of geniuses percolated through the institute in the 1930s. Edwin Hubble, the six-foot-five-inch former boxer turned astronomer, was looking farther into space than anyone before him with his hundred-inch telescope atop Mount Wilson. When not peering through a telescope, he mixed with the Hollywood set, dining with the actor Charlie Chaplin, the writer Aldous Huxley, or the composer Igor Stravinsky. The Nobel Prize–winning geneticist and stringent atheist Thomas Hunt Morgan was developing the chromosome theory of heredity by examining his swarm of mutated *Drosophila* (fruit flies) through a jeweler's loupe. His office, known as the "fly room," was filthy, largely because Morgan had a habit of squashing flies on his desk once he had finished studying them. On the top floor of the aeronautics building, the only location on campus where there was adequate electricity to power his magnetic cloud chamber, Caltech's homegrown Nobel winner Carl Anderson was discovering the existence of the positron—the first empirical evidence of antimatter. Eminent theoretician and president of the Mathematical Association of America, Eric Temple Bell headed the mathematics department. In his youth he had been a ranch hand and mule skinner and although he specialized in number theory during the day, his evenings were spent writing science fiction and detective stories under the pseudonym John Taine.

A younger generation of scientists also held positions at Caltech. Linus Pauling, full professor at only thirty, was shaking up the chemistry department with his revolutionary papers on the quantum mechanics of the chemical bond—the way atoms link up to form molecules—while the shy Charles Richter was working on the earthquake scale that would forever be linked to his name. The bohemian physicist J. Robert Oppenheimer, a master of Sanskrit, medieval French poetry, and a hundred other disparate topics, hurried between Caltech and the University of California at Berkeley, cultivating his mythical image as the poet of protons. Then there were the visiting professors. Niels Bohr,

Max Born, and Werner Heisenberg, the holy trinity of quantum mechanics, all came to lecture.

Caltech students were the beneficiaries of this brain glut. But the students were as famed for their practical jokes as for their intelligence. A Model T Ford materialized on the roof of Caltech's administration building one night and then, as the authorities were trying to work out how to remove it, mysteriously disappeared a week later. Doorways to classrooms could be found bricked up and dorm rooms were routinely emptied of all furniture and reassembled in the courtyards outside. When one of Caltech's professors tried to impose on the students a strict coat and tie dress code for evening meals, the students turned up en masse in coats and ties but without pants. Walking around campus, clutching slide rules in their hands, the all-male student body had a cocksure manner about them, "clearly convinced that Caltech was the greatest and most demanding college in the world and that they, its graduates, must be the smartest students."

Caltech had become a scientific Athens among the orange groves. Most remarkable was the speed with which it had gained this status. A local man like Parsons would have watched the institute transform itself from a small-town technical college into a world-renowned institution. Caltech had been formed by the missionary zeal of George Ellery Hale, a professor of astrophysics and an indefatigable fund-raiser, who in 1903 had a life-changing experience on Mount Wilson, overlooking the city of Pasadena. Hale had grown up in a wealthy Chicago family, haunting the local observatory and keenly reading the classics, which he would later state "helped greatly to arouse [his] imagination and prepare [him] for scientific research." Hale believed that astronomy was a discipline akin to theology and philosophy. A mechanics genius, he was an enthusiastic builder of telescopes and had already invented the spectroheliograph—an instrument for the photographing of phenomena in the solar atmosphere—before he arrived in Pasadena at the age of thirty-five. While lying on the peak of Mount Wilson, Hale became convinced that this would be the perfect location for a new observatory, one which would forever change man's perception of the universe. Like other newcomers to California, Hale was intent on translating his dreams into reality. By 1908 he had conjured up the funding and built a sixty-inch reflector telescope on the exact spot on which he had lain. In a reflector telescope a large concave mirror serves in place of a lens, gathering and focusing the light of the galaxy onto its silver-coated surface. The larger the mirror, the more the viewer can see. The mirror of Hale's telescope was eight inches thick and weighed just under a ton, making it the largest in the world. Along with the 150 tons of steel that made up its intricate mounting, it had been hauled to the mountaintop by man and mule, a task which took over a year to complete. Just below the telescope Hale had built lodging for astronomers hardy enough to make the nine-mile hike along a winding dusty footpath to the mile-high peak. Known as "The Monastery," the building was embellished with mystic Egyptian symbols. On the opening night of the facility, Hale and his astronomical colleagues dedicated the building with pseudo-monastic rituals. Women were not allowed entry.

These ceremonies and rules all spoke of Hale's interest in the mythical aspects of astronomy. The oldest of the sciences, astronomy provided Hale with a link back through history to the ancient schools of Mesopotamia, Babylon, and Greece. Indeed, Hale often described himself as a "sun-worshipper" and his quasi-religious mission to further his science was well known. The *New York Times* called him "Priest of the Sun" and "the Zoroaster of our time." Just as Parsons' quest for the stars in many ways symbolized a deeper longing for some kind of spiritual enlightenment, so Hale was endlessly fascinated by astronomy's links to an ancient world in which the Sun was revered as a god.

Even as the sixty-inch telescope was being completed, Hale had already begun thinking of bigger and better telescopes. On the very day the sixty-inch mirror was set in place on Mount Wilson, a hundred-inch glass disk arrived in Pasadena for grinding and polishing, a process that would take some nine years. When it was eventually finished in 1917, this new telescope became the most powerful in the world, gathering 160,000 times as much light as the eye and increasing the size of the observable universe by a factor of 300 percent. Its power was such that it could detect the light of a candle burning 5,000 miles away.

Yet Hale was not content with just building telescopes. Since he had arrived in Pasadena in 1903, he had been planning a grand scheme: He wished to combine the unparalleled research facilities of his observatory with a nationally ranked school of science and technology. There was no such school in Pasadena, of course, so with characteristic verve he decided to make one from scratch. "Under such conditions, and with the advantages afforded by climate, by the immediate neighborhood of mountains where water-power can be developed and experimental transmission lines installed, who can deny there is a place for a technical school of the highest class?" It was to be an MIT of the West, and it was to be based around the little-known Throop Polytechnic Institute.

Throop was a small engineering school, founded in 1891, that lacked scholarly credentials and had only a handful of students. For Hale it was a blank canvas. He was fortunate to find that one of the school's trustees was Arthur Fleming, the Canadian-born lumber tycoon who had taken the ailing Throop under his philanthropic wing, covering the school's annual deficit out of his own pocket on more than one occasion. In 1915 Hale persuaded Fleming that Throop should be completely overhauled, beginning with the purchase of a site for the new school campus. Fleming immediately bought twenty-two acres of land for $50,000. Hale then began his efforts to attract the brightest and the best to this nascent college. Through letters and over dinners, through proxies or in person, Hale cajoled and flattered the great scientists of the day. When his social astuteness failed him,

he simply bribed them with equipment and the unparalleled research facilities provided by Fleming. He was unflagging in his
enthusiasm and persistence until he had exhausted his targets
into capitulation. "He is the most *restless* flea on the American
continent," confided one target to his wife; "more things are *eating him* than I could tell you about in an hour."

The work took its toll on Hale. He suffered a number of
nervous breakdowns as he rushed across the country courting
support. However, given the gentle climate of Pasadena, the su-
perlative observatory on Mount Wilson, the promise of a break
from the stuffiness of the East Coast Ivy League universities, and
the seemingly bottomless pockets of Fleming, it was not long
before the scientists came. First and foremost among those he
recruited were Arthur Noyes, the preeminent chemist in the
United States, and Robert Millikan, nationally renowned for his
work on cosmic rays and a 1923 Nobel laureate. Once these
two also began searching for suitably brilliant staff, Caltech, as
Throop was renamed in 1920, welcomed more and more greats
into its secluded white buildings.

Hale may have started the process, but the bulk of the credit
for amassing this vast array of talent goes to Millikan, who
swiftly became the doge of the institute. Albert Einstein noted
that as chairman of Caltech's executive committee, Millikan
presided over his faculty like a god. Never one to court popularity, Millikan became a favorite nemesis of the student body. Beneath a giant piece of religious graffiti that read, "Jesus Saves,"
one wit had written, "... but Millikan gets credit." Nevertheless, in 1930 he pulled off one of the greatest coups of his long
and illustrious career by tempting the presiding genius of the
twentieth century to become a visiting fellow at the university.

As paterfamilias to the new scientific race, Albert Einstein
was the final addition to Caltech's roll of honor. One of the reasons Einstein came west was the warm winters, like the many
other immigrants swarming into California each week. But,
more importantly, he was interested in the news that Caltech's

Edwin Hubble had seen, through the telescope on Mount Wilson, distant galaxies streaking away from earth at the speed of light. Here was proof that the universe was expanding, an observation that directly refuted Einstein's view of the universe as a fixed and constant sphere. Intrigued, he and his wife traveled west to see the evidence. Thus for three winters in the early 1930s, Einstein could be found happily riding his bicycle through the campus of Caltech. The addition of this latest and brightest star to the increasingly spectacular constellation of scientific greats in Pasadena meant that Caltech's future was assured.

In 1935, the twenty-year-old Jack Parsons and the twenty-two-year-old Ed Forman wandered into this enclave of scientific brilliance. They had been attracted to Caltech by a newspaper article that had appeared in the *Los Angeles Times* on March 28, 1935. Among headlines speaking of a giant dust storm stretching across New Mexico and Texas and of ominous anti-Lithuanian demonstrations in Germany, one headline seemed to speak directly to them: ROCKET PLANE VISUALIZED FLYING 1200 MILES HOUR. A graduate student in aeronautics (the study of motion in air) named William Bollay had presented a paper at Caltech on the recent work of a member of the amateur Austrian Society for Rocket Technology, Eugen Sänger. Sänger, who had conducted most of his work apart from the scientific establishment and without any outside funding, spoke in strongly optimistic terms about the possibility of rockets being used to power aircraft. What the newspapers were most excited about was the mention of maximum velocities and breaching of the stratosphere.

The fastest a propeller-driven aircraft was able to fly at this point was 440 miles per hour. For Parsons, currently locked in stalemate with his own experiments, Bollay's theoretical paper came as a sign from the scientific heavens. Parsons was by no means awed by Caltech's reputation. He had, after all, been taught by the institute's graduate students while at the adjacent University School. To those living in the small city of Pasadena,

Caltech felt as much a public institution as the city library. Thus, on the very day they read about the lecture, Parsons and Forman strode onto the Caltech campus to find Bollay.

When they finally found his study, they discovered that Bollay was busy on research unrelated to rockets and had no time to help them in their enquiries. But whether because of their earnestness or their considerable and rare breadth of knowledge on the subject, he did not dismiss them out of hand. Instead he directed them to a graduate student named Frank Malina who had similarly mentioned his own interest in rockets to Bollay. It was to be the most serendipitous meeting of Parsons' professional life.

Frank Malina was twenty-two years old, a thin, soft-spoken man, serious and intelligent. He had attended Bollay's lecture the previous day and it had excited him just as it had Parsons. Malina had been born in Brenham, Texas, the son of Czech immigrants. Both his parents were professional musicians and Malina soon developed into a skilled piano and trumpet player. When he was seven years old Malina's family moved back to Czechoslovakia and Malina became increasingly interested in balloons and aircraft, swiftly progressing to rockets upon reading Jules Verne's *From the Earth to the Moon* (apparently the urtext of rocketry). When he returned to Texas five years later, Malina decided he would forgo the musical career his parents had expected of him and instead become an engineer. At the age of seventeen he enrolled at Texas A&M to study mechanical engineering and while he still played music to pay for his expenses—becoming the college's bugler in the process—it was outer space that consistently occupied his mind. In his last year at college he wrote an essay on interplanetary spaceflight filled with the thoughtful speculation of the space enthusiast.

Now that man has conquered travel through the air his imagination has turned to interplanetary travel. Many prominent scientists of today say that travel through space to the Moon or to Mars is impossible. Others say "What man can imagine, he can do."

The paper was infused with a spirit of human endeavor, with a belief in science's ability to better mankind. It was characteristic of Malina's scientific approach—humanitarian and levelheaded.

When Parsons and Forman approached him, his feelings on rocketry had been dampened by more prosaic academic work. But he recognized an opportunity in their query. Following an endowment by the Daniel Guggenheim Fund for the Promotion of Aeronautics, Caltech had gained an aeronautical laboratory on campus. Known as the Guggenheim Aeronautical Laboratory at the California Institute of Technology (GALCIT), it had a wind tunnel in which Malina was working. The tunnel, ten feet in diameter, lay at the center of the GALCIT building. Offices, laboratories, and classrooms were distributed haphazardly around the tunnel as if it were a magnet to which study had been irresistibly drawn. The reason for its attraction was the tunnel's ability to generate wind speeds of up to 200 miles per hour, in which scale models of newly designed aircraft could be tested for lift, drag, and stability. All the aviation companies of the West Coast used it, as did many others from further afield, each paying $200 per hour for the privilege. Since it was also used for GALCIT's own research projects, the tunnel was always busy. For Malina, long hours spent observing models in the tunnel had only demonstrated the limitations of propeller-powered airplanes. While employment in the tunnel had its benefits—on hot days he could lie on his back and smoke cigarettes as a cool breeze played over him—his wages, which stood at a trifling 25 cents an hour, heightened his frustration with the other shortcomings of his work.

A shared dream can unite even the most disparate characters. When Parsons and Malina met, their enthusiasm for rocketry and space travel bonded them like a Masonic handshake. They also shared an intellectual affinity. Malina, the skilled musician and artist, and the well-read and cultured Parsons soon found they shared many other interests. They were both interested in classical music and politics and, because of their respective travels in Europe, were concerned with the rise of Fascism there. They even shared the same birthday. As for the difference between their scientific educations, this was somewhat beside the point: Rocketry was not even acknowledged in any formal science courses. The topic of academic credentials was never discussed. "It seems to me that at most he had finished high school," remembered Malina. "When I met him he already had a certain amount of experience with the manufacture of explosives . . . but I think what was outstanding about him was that he was not of any fixed view on which way to go . . . He had a very flexible sort of attitude." With Forman they formed a complementary triumvirate: Parsons would act as the chemist, Malina the mathematician, and Forman the engineer. All were aware of Robert Goddard's martyrdom at the hands of the establishment and of the general scorn in which rocketry was held by the American academic elite. Surrounded by the greatest scientific minds of the time, Parsons, Forman, and Malina must have felt, conversely, as if they had fallen in amidst the barbarian hordes.

However, if their study of rocketry was to stretch beyond enthusiasts' banter, the blessing of Caltech was needed. Here was a treasure trove of chemicals, tools, and engineering know-how. If they could persuade the university authorities to allow them to use the facilities, and perhaps to give them some funding as well, their work—to study, create, and fly rockets—could begin in earnest.

As they began putting together a proposal, the three recognized the need to avoid being tainted by the "moon-rocket" tag that had plagued Goddard so cruelly. Even the American Interplanetary Society had bowed to the derision that talk of space travel caused and had changed its name to the more modest American Rocket Society (ARS). In shooting for the stars, they would first have to lower their gaze.

Parsons and Forman came up with a moderate description of their aims for the Caltech authorities: "The design of a high-altitude sounding rocket propelled by either a solid or liquid propellant rocket engine." Their proposal did not sound outlandish. They planned to build a research rocket that could carry meteorological instruments or a camera into the ionosphere, twenty-five miles up, and then return to earth by parachute, with information on the atmosphere. The fact that it would be able to travel higher than balloons might even tempt the meteorological department of Caltech to take a financial interest in it. But they all knew that no known rocket technology had come even close to achieving such heights. There had been much theorizing about the possibility of 1,200-miles-per-hour rocket planes, but no actual experiments, even among the VfR, had approached these goals. There had been successes: The VfR had gotten one of their rockets to fly to the height of one kilometer in the early 1930s. Although cloaked in secrecy, the Russian rocket societies had also attained heights of a few hundred meters. The British Interplanetary Society was not so lucky. The Explosives Act of 1875 forbade them from any actual experiments with explosives. However, in the United States the auspices weren't much more promising. The great Robert Goddard had only managed to get his liquid-fuelled rockets to soar to a height of ninety feet before his exile.

Malina, preferring to put theory before experimentation, suggested that the three make an even less ambitious proposal. Nothing could be allowed to conjure up the bugbear of serious science—sensationalism. He suggested a two-part program: theoretical studies on the thermodynamics and flight performance of a rocket, then the creation of a working rocket motor (the chamber in which the fuel combusts). Malina argued that they should not start their project by designing a whole rocket as Parsons had wanted, complete with a space-age shell, launching tower, parachute, and the like. Malina also suggested that only static tests should take place. In a static test the rocket motor is inverted on a stand. With the exhaust nozzle aimed skyward, thrust can be measured by the push of the motor on pressure gauges. The rocket wouldn't move much, but it could be examined easily. Until the three of them fully explained to Caltech and to themselves the intricacies of how a rocket functioned, Malina said the extent of their initial plan should be "a series of theoretical studies which address the thermodynamic problems of the reaction principles and the flight performance requirements of a sounding rocket." It was back to basics for all of them.

Parsons and Forman were shocked by the austerity of Malina's plan. They had wanted to continue their own experiments at Caltech, not begin the whole project again. The very idea of static tests seemed contrary to the work they had been pursuing for the last eight years of their lives; the damn motor wasn't even pointing in the right direction! An addiction to a rocket's blastoff and the subsequent compulsion to build a bigger and better rocket were endemic among the early rocket enthusiasts. To spend months, possibly years, constructing a rocket motor that wasn't even going to get off the ground was the equivalent of giving milk and cookies to a junkie.

Arguments flared, Parsons and Forman threatened to leave, and the embryonic group teetered on the brink of dissolution. But they needed each other. Malina was Parsons' link to Caltech and scientific respectability, not to mention the institute's vast resources; Parsons in turn was Malina's connection to the scattered history of rocket experiments and a storehouse of valuable firsthand experience of constructing and testing rockets. Each served as the foil for the other: Malina trying to calm Parsons and introduce him to basic scientific method, Parsons spurring Malina on to experiment, experiment, experiment. The argument over the rocket proposal anticipated the creative, and destructive, tension that would characterize the men's relationship in the years to come—Parsons and Forman, the impetuous experimenters, versus Malina, the cautious theoretician.

Eventually, Parsons and Forman grudgingly downgraded their project to the construction of "a workable engine with a reasonable specific impulse" ("specific impulse" being the measure of a motor's efficiency in providing thrust to a rocket vehicle). It may have lacked the explosive payoff of the rocket launch, but this new project was by no means simple. To make a working rocket motor they had to thoroughly understand every part of it. The correct size of the combustion chamber needed to be ascertained: a larger chamber allows more fuel to be burned, making the rocket more powerful but also heavier, while a smaller chamber might be lighter—but could it produce enough thrust to lift itself? The best fuel mixture—solid or liquid—had to be decided upon; the liquid fuels were more powerful but more dangerous, while the solid fuels provided less thrust but were safer. The group would be forging ahead into largely unexplored territory. Malina would handle the mathematics side, using formulae to determine the pressure of a burning gas inside a rocket combustion chamber or the theoretical burn times for differing types of fuel, factors crucial to studying and improving the thrust. Parsons would provide the practical and chemical element, creating tailor-made explosives. By now he had a voluminous knowledge of chemistry. Forman would act as the chief mechanic, constructing and designing the metal motors in the workshop.

With Malina acting as the group's intermediary with Caltech, the proposal was put to Clark Millikan, Robert Millikan's son, in charge of the wind tunnel and a professor in the aeronautics department. Millikan had a fearsome reputation. Imperious in the classroom, he taught his students while smoking a cigarette through a long holder and was renowned for giving tests that were so difficult 95 percent of his pupils could never finish them. Students speculated that his draconian attitude stemmed from resentment of forever living in his father's long shadow. Seeking permission to change his thesis to the study of rocket flight and propulsion, Malina showed Millikan the proposal that he and Parsons had worked on and fought over. Millikan turned down the idea immediately. Rocketry was not practical, Millikan told him. In fact, it was a harebrained idea, fit only for Hollywood films and thrill seekers. He suggested that Malina finish his degree and get a job at one of the aeronautical companies that were located in the area. A job in the industry was the expected goal of aeronautics graduates after all. Rockets should be left to the comic books. Clark Millikan's rebuff was a blow to Parsons and Malina, but not an entirely unexpected one. They knew whom to try next—the one mind at Caltech who made a living out of radical thinking, and who just happened to be the director of GALCIT itself.

Robert Millikan had tempted Theodore von Kármán, the most gifted aerodynamicist of the era and head of the aerodynamics institute at the Technical University of Aachen, Germany, to immigrate to the United States and become GALCIT's first director in 1930. Kármán had been a child prodigy in his native Hungary, able to multiply six digit numbers in his head at the age of six. Now fifty-four years old, he was still short in stature and spoke with a strong Hungarian accent, and was much revered throughout the scientific world. His dark, untamed eyebrows hooded a piercing, curious gaze, and he was immediately recognizable hurrying across the Caltech campus, beret on head, cigar in hand, an irrepressible, at times mischievous, presence. He had a reputation as a ladies' man and could often be found romancing his students' wives at college functions. He was also famed throughout the student body for his bursts of absentmindedness. On one occasion he was in the midst of giving a lecture to his class at Caltech when, halfway through it, he realized that he had been talking solely in German. None of his American students had uttered a word in protest, so Kármán ended his sentence and, without missing a beat, continued the lecture in English as if nothing had happened.

He lived with his sister, Pipö, in a house bedecked with hundreds of artifacts collected from their travels across the globe.

At times they seemed to inhabit a dream world as Parsons had as a child. When Kármán and his sister entertained guests, they would often adorn themselves in Japanese kimonos for the engagement. He was a flirt, a respected advisor, and, to his pupils, a revered father figure. Upon arriving at Caltech, Kármán had established a charismatic style of teaching. He formed tight bonds with his students that went far beyond the classroom, bonds that incorporated friendship, mutual respect, and genuine adoration by the pupils. He stressed a more creative approach to learning. For example, he would ask how an electron would "feel" in its environment. For Kármán the imagination was all important in tackling any problem.

The most vivid example of his unorthodox scientific mindset and his frustration with those who lacked vision came when he was called to help explain the Tacoma Narrows Bridge collapse. The $6 million steel suspension bridge, 6,000 feet long and described on its completion in 1940 as the greatest single-span bridge in the world, had broken up in a light gale shortly after completion. It had been filmed bucking, bending, and twisting as if it were made of rubber, before collapsing into Puget Sound. The bridge's engineers had been mystified by the disaster, obsessing over the weight and pressure, the "static forces," acting on the bridge. Kármán, as an aerodynamicist, saw the disaster differently. The problem wasn't the weight on the bridge so much as the aerodynamic forces, the forces of air in motion, working on its frame. He theorized that the bridge had acted like a poorly designed airplane wing and had responded to the lifting forces of the wind flowing across it. Under certain conditions the bridge's design promoted turbulent airflow—which later became known as Kármán vortices—causing the bridge to oscillate with the rhythm of the wind flow. Once this oscillation began, the bridge bucked and swayed; the wind did not even have to be strong to literally bend steel. The bridge's engineers scoffed when Kármán suggested putting a scale model of the bridge in the wind tunnel at GALCIT. But he went ahead with the test and provided positive proof of his theory. The so-called experts' stubborn self-assurance crumbled, and Kármán introduced a whole new way of designing stable structures.

"In desperation," Malina, Parsons, and Forman went to the iconoclastic Kármán to see if he could save their seemingly doomed project. In his autobiography, Kármán remembered being "immediately captivated by the earnestness and the enthusiasm of these young men." Their unusual backgrounds made them all the more fascinating. Parsons may have been a largely self-taught chemist, but Kármán immediately recognized his "considerable innate ability." He saw Parsons as a romantic, a dreamer, "searching for some private gate to happiness," but he felt that in a nascent science an unorthodox mind-set was a valuable commodity to have —just as Malina, while realizing that Parsons "was not mathematically talented," had seen Parsons' "freewheeling brain" as "extremely valuable" to these early stages of development. Parsons displayed exactly the kind of attitude Kármán was famed for endorsing.

Impressed by the boldness of the three young men and amused that Clark Millikan's rejection had not affected their ardor, he mulled their suggestion over for a couple of days before summoning them back to his office. He told them that he could not give them any funds: The depression was cutting into even the institute's vast endowments. However, he would allow them to work under GALCIT's auspices. What's more, they could use the laboratory's equipment after hours, despite the fact that Parsons and Forman had no connection whatsoever with Caltech.

The group was overjoyed. Their rocket project had gained the support of one of the greatest minds of the twentieth century. Now named "The GALCIT Rocket Research Group," they would exist unofficially for the next four years, without an appointed leader or bureaucratic structure. They would rely on trial and error, good luck, and, above all, imagination. By the time their experiments were finished, the study of rockets and space travel would be transformed from a foolish pursuit into a genuine science. For Parsons the project was the culmination of a dream from his childhood—to belong to a group of men who were doing something noble and wonderful, an Arthurian band of adventurers making a quest into space.

Rocketry had not been the only thing on Parsons' mind. As well as gaining backing for his rocketry experiments, the previous years had also seen Parsons finding a wife. Following his eye-opening European trip several years earlier, girls had become a serious priority—albeit one not quite as serious as rocketry. With the rakish Forman he had been wandering the social circles of Pasadena, in particular the dances put on by the various churches in town. In the winter of 1933, the two attended a Christmas dance held at the First Baptist Church. There were prizes for the best dancers, and with the depression still biting, the dance provided some welcome gaiety amidst the rigors of the workweek. As they walked through the church door, Parsons was stopped in his tracks by the sight of a young woman in a pink dress studded with rhinestones being spun down the floor of the hall by her dancing partner. Transfixed by this twirling girl, Parsons insisted that he and Forman introduce themselves. Her name was Helen Northrup; she was twenty-two years old, four years Parsons' senior. Tall—she stood five feet, nine inches—and thin, she had dark brown hair and "Irish blue eyes."

She was immediately enamored of the boys' talk of rockets and their hopes of traveling to the moon. Nevertheless, Helen was not one to be bowled over by Parsons' charms; indeed, recognizing Parsons' attraction to her, she turned her attentions to Forman instead. "Jack saw me to the door that night," remembered Helen, "and I refused to give him any information—my telephone number, or address...I was a teaser if you will." Undeterred, the following day Parsons tracked her down to her house in Pasadena and, along with Forman, was eventually invited in for a game of cards. Helen was intelligent and well-read, and she shared with Parsons a great love of classical music, particularly Stravinsky. The formal visits continued, and Parsons' puppy dog persistence eventually won her over.

Helen's life had been a difficult one. She had grown up in Chicago, along with her two younger sisters. They were the daughters of Thomas Cowley, an Englishman working for the Standard Oil Company, and Olga Nelson, the daughter of Swedish émigrés whose father, as family lore recalled, had once been an adviser to the Russian tsar. During a bitterly cold winter in the early 1920s, Helen's father had contracted pneumonia and died. His death left Olga to care for the three young girls. While working as a desk clerk at a hotel she met Burton Northrup, a traveling salesman. Olga, needing both financial and emotional support, soon married him. But Helen's new stepfather had anything but her best interests in mind. He sexually abused her and her sisters, obsessing over them and refusing to allow them boyfriends or even to speak to boys.

When Helen was twelve years old, the family relocated to Pasadena. The destination had been chosen, remembered Helen, through Olga's use of a Ouija board. So the three daughters, their abusive stepfather, and defeated mother made the long drive west. Upon arriving in Pasadena, the Northrup family grew again when Olga gave birth to two more girls, Sara—known as Betty—and Nancy.

In 1928 Helen's stepfather was convicted of financial fraud and imprisoned for a brief period. A shy and sickly Helen was forced to drop out of high school to support her family. She found employment in Pasadena working as a window dresser for a local dressmaker in town. A talented designer herself, she enjoyed the work, but upon the return of her stepfather she was forced to become a secretary in his debt collection firm. Helen was still forbidden to date boys, although she had her fair share of admirers. Her extracurricular activities were limited to Sunday school and the activities of the Baptist church.

However, Helen's stepfather encouraged her courtship with Jack. Parsons seemed a wealthy, cultured boy, an ideal match for an eldest daughter who was by now of marrying age. Helen was happy, too. Here was a man she was allowed to see, one who was refined, sophisticated, and handsome: "Jack was a well-traveled man worldwide so of course he had a bit of class...He always dressed [well], even when he tested rockets." Indeed Parsons' dress code was remarked upon by everyone. Even in a time when formal dress was common, Parsons' devotion to his suits was unusual. Another friend from the time remembered how Parsons always "wore a regular suit with a vest," and recalled, "I don't think I ever saw him without the vest." In Parsons' mind the suits may well have acted as mementoes of his younger, wealthier days.

Ruth Parsons, Jack's mother with whom he was still living, approved of the match as well, supplying Helen with clothes from her delectable Parisian collections. Helen was quiet and attentive, supportive and organized, and she seemed interested in everything Jack was doing. A typical Sunday morning would see Parsons drive over to Helen's, wait for her to finish the household chores her stepfather made her do, and then spend seven or eight hours with her, often in the company of Ed Forman. The two also spent time alone together, driving up to the nearby San Gabriel Dam and looking down along its canyon, or going into the mountains above Burbank Airport and watching the planes glide in. They traveled outside Pasadena together, to the town of Avalon on Santa Catalina island, and they took astronomy classes together, possibly at the Mount Wilson Observatory. It was while gazing at the stars that Parsons admitted to her, as he later wrote, "I first learned how much I loved you."

Occasionally, however, Parsons' enthusiasm for his rockets suggested an infuriatingly solipsistic character. On one evening when Jack had promised to take Helen on a date, Forman came along, and the men quickly became so engrossed in their talk of new rocket designs that Helen was hardly noticed. "I was completely ignored," she recalled, "so I walked home. Hours later [they said to one another] 'Oh, where's Helen?'" The practical and social necessities of life could often seem very distant to Parsons, especially when he was occupied with rocketry. "Jack was already erratic," Helen would later recall of this time. "His mind wasn't on the 'now'."

By July 1934 the nineteen-year-old Parsons had proposed marriage. The three-carat diamond engagement ring that adorned Helen's finger was that which Parsons' absent father had given to his mother. Along with the ring, Parsons also gave his wife a slightly less romantic gift—a .25-caliber pistol. Helen later recalled: "He had given me a nice diamond that he thought should be protected. I think I would have been more scared using it than losing my ring." The couple, who still lived with their families, planned to marry within a year. However, money was needed if they were to have their own home, and no sooner had they become engaged than Parsons went to work at the Hercules explosives plant in northern California. Separated by five hundred miles, Parsons wrote a letter a day to his bride-to-be in Pasadena. He would tell her everything about his day. For example, he had torn down the naked pinups on the walls of his dorm room and replaced them with a single picture of Helen. Also, he was trying to "keep up appearances" and grow a moustache for her but was failing to do either.

Within a day of arriving he longed to be back with Helen, writing her, "Your ring on my finger is like the Touch of your hand—and I know your heart and thoughts are with me." Two days later he was lamenting his fate, "Isn't it sickening to love each other as much as we do and then be parted? I miss everything about you—the sound of your voice—your lips—your little ways."

Parsons rhapsodized on both their love and his own scientific longings, conjoining the two into a grand romantic quest.

"If the night is clear when you get this letter," he wrote, "go out and look at the pole star—the pointer in the handle of the dipper. Let us make that our star. It is the star of the abiding—abiding as our love is steadfast—pointing us the path to the skies. Nights may be long and skies cloudy but shining above all clouds and parting our star remains, shining clear and bright above, undimmed by time or any other of the little things of earth. Let it symbolize our love—'Ad Astra Per Aspera'—the stars are our goal and nothing in the outside of the galaxy will keep us from it."

The longer he was away from Helen, the more lyrical and ecstatic his letters became, filled with allusions to his favorite mythologies and myths, to Valhalla, "the star palace," and El Dorado, "our city of gold." The Avalon of Arthurian mythology soon grew indistinguishable from the Avalon of Santa Catalina, under whose stars they had also walked. Unable to afford long distance phone calls, Helen and Parsons even tried to communicate telepathically with one another, "tuning in," as Parsons called it, at nine o'clock each night. "I think I made it," Parsons wrote to Helen after one occasion; "I was so near to you—I could hear your voice—it wasn't clear ... I saw us in Avalon—walking down that street from the house hand in hand ... Perhaps it's some trick of loneliness and imagination—but somehow I doubt that." He signed his letters "Jack," with the *J* drawn as a flying rocket and beneath it the motto, Semper Fidelis—"Always Faithful."

The wedding took place on April 26, 1935, at the Little Church of the Flowers in Forest Lawn Memorial Park in Glendale, Los Angeles. The two families—one numerous and riven by abuse, the other tiny and claustrophobically close—saw Jack and Helen pledge their marriage vows. But even on an occasion as important this, Parsons was in something of a reverie. At the end of the ceremony, to Helen's dismay, he forgot to kiss her. "I turned my face to him and nothing happened." The couple took a brief honeymoon at the Hotel del Coronado in San Diego, where they spent the night dancing in the ballroom. Parsons was only twenty years old; Helen was twenty-four.

Back in Pasadena the couple moved into a new house together on Terrace Drive. In order to pay for their house and remain close to his new wife, Parsons began working at the Halifax Powder Company, a manufacturer of explosives, flares, fireworks, and small munitions in Saugus, a town thirty miles northwest of Pasadena in the hills above the San Fernando Valley. But almost all of the money he earned from this job was being pumped into the Rocket Research Group. "I knew his first wife rather well," recalled Parsons' friend Robert Rypinski, who used to visit the couple regularly, "and she used to cry on my shoulder because the only thing Jack would save money for was to go out and shoot rockets. And they lived in rather meager quarters and he wouldn't spend any money on clothes for her, decent cars or anything. He just wanted to work on rockets."

His manner of dealing with explosives also caused her consternation. On one occasion Helen joined Parsons and Forman on one of their recreational skyrocket launching trips in the desert. Sitting in the back seat of the car, she lifted up a rug covering the floor to find it had been hiding sticks and sticks of dynamite, no doubt taken from Halifax by Parsons. Nervously leaning forward to the front seat where Parsons and Forman were sitting, she asked whether the explosives were safe. As the truck bumped heavily along the desert road, Parsons turned to her with an amused grin and told her not to worry: "The detonator's in the front seat." In their new house, Parsons insisted on constructing a home laboratory on the porch so that he could work on his chemicals and explosives at any time of the day or night. "You should have seen the state of his [home] laboratories," remembered Helen. "They were dangerous places." On one occasion Parsons' grandmother was rushed to the hospital at the very moment that Parsons was heating a large vat of explosive on an incinerator he had constructed in their backyard.

Handing a large spatula to Helen, Parsons told her to keep stir-ring and under no circumstance to stop until he had returned from the hospital. With that he left a petrified Helen to stir the foul-smelling chemicals until his return some hours later with the news that his grandmother would be all right. Life with Par-sons, remembered Helen, was "happy but haphazard."

Even Frank Malina felt rather uncomfortable when he saw Parsons' setup. Large barrels of gunpowder, open to the ele-ments, could be found on the porch, not to mention a signifi-cant accumulation of tetranitromethane—a highly unstable organic compound, briefly considered by Parsons as a rocket fuel—in the kitchen. Despite the hazards, Malina visited often, as did Ed Forman. The men often talked about the rocket proj-ect, but their conversation ranged far and wide. They argued about politics late into the night, about the Spanish Civil War and the alarming rise of Fascism in Europe. The storm clouds of war were gathering, and both Parsons and Malina knew with the certainty of youth that change was needed. The enthusiastic socialist sentiments that they shared were now common not just throughout America's intellectual circles but throughout the social sphere, thanks to the successes of President Roosevelt's New Deal reforms. However, the particular strain of left-wing thought that characterized the rocket group can be seen in one of Malina's letters home to Texas from 1936:

> Events in Europe are certainly heading to another war. There seems to be only one hope; overthrowing of the cap-italistic system in all countries and an economic union of nations. The American Legion, Hoover and his cohorts and other patriotic organizations now extolling "Americanism" seem to show that in the US people are beginning to doubt the reasonableness of the capitalistic system.

As the evenings stretched into nights, Parsons would begin to read aloud from his favorite poets. "He probably thought I was a bit of a square, in that I have never been very sensitive to poetry," remembered Malina, "and he would show me exotic poems, and I would look at them, but they didn't make any sense to me." Indeed, Helen recalled Malina as, "studious, very studious. Ed, Jack and I would try to loosen him up a bit." Forman would never quite jell with Malina in the way that Parsons did, and later in his life he would become convinced that Malina looked down upon him because of his lack of a university education. But Malina never suggested that this was the case. For now, however, the group was pulled together by a mix of bonhomie and rocketry. Many drinks would be consumed on these evenings, the famed "Parsons Poison Punch" or maybe some of the absinthe Parsons had been concocting in his home laboratory. The three men and Helen also tried marijuana in yet another groundbreaking experiment for the GALCIT Rocket Research Group.

Malina would often play piano duets with Helen, or the group would play records deafeningly loud on the much-used phonograph. They bought cheap tickets for classical concerts—the civic auditorium in Pasadena put on performances of *Il Travatore* and *Madame Butterfly*—and during the breaks in the performance, they would work their way to the empty seats in the front and watch the remainder of the concert in style. Over the next few years Malina and the Parsons would also take trips to the beach and into the mountains together. Writing to his parents, Malina declared, "I have found in Parsons and his wife a pair of good intelligent friends."

Throughout these early years of marriage, Helen and Jack enjoyed moments of great affection together. When the cats wailed at night, the couple would climb out of bed and go and stand outside, naked in the moonlight, to listen to them. Inside the house they also kept canaries and adopted any stray animal that happened upon them. During a storm in the area, one of the trees in their garden fell down. When Parsons dragged it into the house to use for firewood, he inadvertently brought along a small owl that had been nesting in it. The owl soon became domesticated and took to perching on the mantle, further expanding the menagerie.

Parsons' generosity to his human houseguests was also well known. Too good-natured to ever get angry with anyone, he often left it to Helen to read the riot act. An example in point was Tom Rose, a vastly experienced explosives expert who had worked at the Hercules Powder Company for years and had taught Parsons many of the tricks of the trade. When he asked if he could stay at the house for a few days, Parsons and Helen gladly let him. As the weeks passed and Rose showed no sign of moving, Helen grew increasingly frustrated. But Parsons could not bring himself to tell Rose to leave. His profound inability to say no to people would be a habit others would unscrupulously exploit later in his life.

4

· · · · · · · · · · · ·

THE SUICIDE SQUAD

*New ideas come into this world somewhat like falling
meteors, with a flash and an explosion, and perhaps
somebody's castle-roof perforated.*

—HENRY DAVID THOREAU

In April 1936, as Parsons and his colleagues celebrated their acceptance into the Caltech fold, a two-hundred-inch reflector disk arrived at the Pasadena train station on its way to a new observatory at Mount Palomar, fifty miles to the south. Costing $300,000 and weighing nearly twenty tons, it was wrapped in felt, cushioned with sponge rubber, and packaged in a steel-plated crate. Crowds gathered in the street to wonder at this testament to engineering and to guess at what it might reveal. For this glass eye, twice as powerful as Hale's telescope on Mount Wilson, promised to bring into view 1,000 million galaxies, each containing some 100,000 million planetary systems.

The more worlds that were found, the more intently the rocket societies operating across the globe sought to reach them. In Nazi Germany, however, the most successful and voluble group of amateur rocketeers, the VfR, had suddenly fallen silent. Letters written to them were not answered, and nothing was heard of their recent developments. Indeed, the word *rocket* was banned from German newspapers in 1934. The Versailles Treaty, which severely restricted Germany's armaments in the wake of the First World War, had conspicuously left out any mention of rockets. The members of the VfR were now being pressured into working for the military on a secret rocket project. Some did not need much persuading: "It seemed that the funds and facilities of the Army were the only practicable approach to space travel," wrote Wernher von Braun. This was reasoning that any rocketeer bitten by the space bug and starved of funding might well understand, particularly as the VfR was far ahead of any other group in terms of rocket progress. Unfortunately, the German army did not share in such intergalactic enthusiasms. "The value of the sixth decimal place in the calculation of a trajectory to Venus," recalled one army officer, "interested us as little as the problem of heating and air regeneration in the pressurized cabin of a Mars ship." The army wanted rockets able to carry explosives not astronauts.

Writing to his counterpart in the American Rocket Society, P. E. Cleator, the founder of the British Interplanetary Society, expected the worst: "Apparently there is some trouble brewing in Germany—trouble about which Herr Ley [Willy Ley] dare not write . . . It seems to me that rocket research in Germany is becoming a closed book—until the fighting begins." No one in the United States seemed to pay the VfR's fate much notice. Rocketry was still a taboo subject in government as in the universities. Perhaps if Robert Goddard had not been exiled in New Mexico, he might have reminded the powers-that-be of a prediction of his: "I would not be surprised if it were only a matter of time before the [rocket] research would become something in the nature of a race." The rocket was going back to war.

Once Kármán had given the GALCIT rocketeers his blessing, their infinite dreams were swiftly diffused by a flood of particulars. What rocket to build? What fuel to use? How best to measure the rocket's performance? They eventually decided to build a rocket motor to be propelled by the highly volatile combination of gaseous oxygen and methyl alcohol (methanol), a mixture first suggested by Eugen Sänger. At the end of the school day, as the Caltech students walked back across campus to their boarding houses, Parsons, Forman, and Malina could be seen heading against the tide to the now empty machine shop to begin shaping their rocket motor.

It was a time of penny-pinching and pilfering. With no official funding they had to pay for every bit of material they needed out of their own pockets. They scoured junkyards for tube ends and pressure gauges and stripped old ovens bare for dials and piping. Whole weekends were spent driving around Los Angeles and Long Beach looking for high-pressure tanks and meters. Often they would have no luck at all, and their spirits would sink. "Two instruments we need cost $60 a piece and we are trying to find them second hand," wrote Malina to his parents. "I am convinced it is a hopeless task." Preparations were hampered by their other work. Parsons and Forman were still working at the Halifax Powder company during the week and meeting with Malina in the evenings or on weekends. As well as working all day in the wind tunnel, Malina, a skillful artist, was illustrating one of Kármán's textbooks.

As Parsons taught Malina about the practical elements of firing a rocket, Malina was schooling Parsons in scientific procedure—taking notes, drawing tables, writing detailed chemical analyses of the different fuels involved. The two continued to struggle; Parsons did not take well to the usual scientific deductive process, preferring instead to use his intuition. "We were constantly putting his nose down to the paper to make him show us what were the alternatives," remembered Malina.

When summer arrived, Parsons and Forman alleviated the strict regime of scrimp-and-save by launching some of their old black-powder skyrockets, just for the thrill of it. The serious-minded Malina was not entirely comfortable with his colleagues' lack of discipline, but he understood their restlessness. "Their attitude is symptomatic of the anxiety of pioneers of new technological developments," he wrote dryly. But the launching of these rockets was also a release for Parsons; it kept his passion from being reduced to labor.

In August 1936, the Rocket Research Group learned with some excitement that the hidden god of rocketry, Robert Goddard, was leaving his self-imposed exile in New Mexico to visit California and Caltech at the suggestion of his benefactor, Harry Guggenheim. Goddard had just released a report, sponsored by the Smithsonian Institution in Washington, D.C., entitled *Liquid-Propellant Rocket Development*. In it he spoke of the research he had conducted since the humiliation of 1919. It was blandly written, as if designed to douse enthusiasm, and when in guarded terms he revealed the staggering news that he had, in 1926, launched the world's first liquid-fuel rocket flight (the rocket had traveled 56 meters at approximately 60 miles per hour), he subsumed any excitement he might have felt in caution and suspicion: "I am rather reluctant to specify what heights I believe possible." Nevertheless, for Parsons and other rocketeers his report was a revelation. Not only had Goddard hinted that he was well ahead of anyone else in the rocketry world, but his position as a responsible physicist backed by a national organization argued that rocketry should be taken seriously.

Goddard's meeting with Robert Millikan did not go well. Thinking of the coup it would be for his university, Millikan tried to tempt Goddard to move his research work from the desert to Pasadena and indulge in a little teamwork with Millikan's pantheon of scientific greats. Goddard turned him down point blank. Millikan did not push the matter, but he did ask Goddard if he would meet Frank Malina, who was working on what Millikan was now calling the Institute's own fledgling rocket project. The encounter was brief, but Malina managed to engineer an invitation to visit Goddard in his laboratory in Roswell, New Mexico.

When Parsons heard the news, he was thrilled. He drew up a list of technical questions he hoped Goddard would answer for him. "What jet velocities, efficiency, and energies did he get from solid fuels in steel chambers with Taper nozzles? . . . What was the most powerful and promising powder fuel which he investigated, and how did it compare with the liquid fuels, irrespective of difficulties of fuel injection? . . . Did he ever investigate nitrogen pentoxide, $N2O5$, or any other oxidiser besides liq. Oxygen?" But from the moment Malina arrived in the blistering heat of Roswell, the suspicious Goddard kept him at arm's length. Goddard showed Malina the test stand, but there was no rocket attached. As they toured the workshops, the components of Goddard's rockets stayed under tarpaulin wraps. When Malina asked about particulars, Goddard switched the talk to generalizations. Finally, as Malina's frustration became visible, Goddard showed him a pristine newspaper clipping. It was the damning *New York Times* editorial that had led to his self-imposed seclusion sixteen years earlier. Goddard still "appeared to suffer keenly" from its criticism. It was obvious to Malina that Goddard would never help the rocketeers. Conciliatory, Goddard suggested that Malina come and work for him when he had finished his studies at Caltech. By that time, however, Malina would have no need for him.

Though he had laid the foundations for the science of rocketry, the solitary Goddard was on the verge of being surpassed by groups of workers who could work faster and more flexibly than he could on his own. While Goddard was hamstrung by having to attend to every aspect of his rockets himself, the Rocket Research Group was already replicating, in miniature, the systematic division of labor that would characterize all later developments in rocketry throughout the world: Malina dealt with theoretical problems, Parsons and Forman with the practical experimentation. Goddard's genius cannot be denied. He was, for many years, the only American of any academic standing to push forward the idea of rocketry as a science. In its early days the Rocket Research Group (and the American Interplanetary Society before it) would continue to refer to and reenact his experiments as they sought to grasp the finer details of the science. But despite the fact that his work anticipated much of the technical detail of, among other things, the German V-2 rockets used in the Second World War, Goddard's unwillingness to communicate his findings with anyone else ultimately kept him from playing a much more important role in the history of rocketry. His failure to cooperate with others frustrated even the easygoing Theodore von Kármán who, in 1967, summed up Goddard's achievements: "There is no direct line from Goddard to present-day rocketry. He is on a branch that died. He was an inventive man and had a good scientific foundation, but he was not a creator of science, and he took himself too seriously."

On October 2, 1936, Parsons and Malina held a party to celebrate their twenty-second and twenty-fourth birthdays. They were also celebrating the near completion of their first rocket motor after six months of designing and building. The first test would take place on October 31—Halloween. Although both Malina and Kármán stressed the importance of theory over experimentation, hard facts were now needed. The rocket motor tests would provide data that would help the team plot thrust-time curves, understand fuel consumption speeds, and discover what temperatures and pressures were occurring within the motor during the process. The tests would teach them the best way of injecting fuel into the motor, what the best shape of the rocket motor would be, and what type of exhaust nozzle would yield the highest performance.

Since no one knew quite how noisy, or dangerous, the tests might be, the rocketeers wanted to find a spot somewhere off campus. Parsons knew exactly the place—the still-wild northern reaches of the Arroyo Seco, near the ominously titled Devil's Gate Dam.

The group had recently attracted a new member as word of the unusual project filtered through the university. Apollo M. O. Smith, known by all as Amo, had just begun work on his MA at GALCIT. His unusual first name had come from his father who, obsessed with the classics and piqued by his unexceptional surname, had named all his children after characters from the Greek myths. Thus Apollo grew up in a household with sisters Diana and Athena, brothers Hermes and Orpheus, and the family dog, Cerberus. Smith flew gliders at high school and spent some time working on oil rigs as an adult. At Caltech he gained a reputation for eccentricity by wearing a pith helmet to which he attached a ventilator and a weather vane so that his head stayed cool. "It wasn't really my main interest," he said of the rocket project, "but it sounded like fun." He brought with him his refinery experience of pipelines and inflammables.

The day before the Halloween test, Parsons, Forman, and Malina spent hours driving back and forth between Los Angeles and Pasadena, picking up the pressure tanks, fittings, and instruments necessary for the experiment. By 3:30 on Saturday morning they were exhausted. They retired home to grab three hours of sleep and at dawn made their way back to Caltech to put the finishing touches to the rocket apparatus. Malina commandeered an institute truck to transport the apparatus, complete with one large tank full of methyl alcohol and one of gaseous oxygen. By 9:00 A.M. Apollo Smith had joined them, as had Rudolf Schott, another Caltech graduate student, who had offered his services for the day if he could watch the experiment.

It was a thirty-minute ride into the wilderness along a road that soon turned into a dirt track. Parsons stood on the back of the truck, steadying the extremely explosive cargo as it bumped up and down on the uneven road. Just behind the Devil's Gate Dam, they spied the perfect spot on the dry riverbed below. With the truck perched precariously over the valley edge, they began the backbreaking work of lugging the tanks, pipes, and other equipment down to the valley floor. They filled sandbags with dirt and piled them high as a protective barrier. All the screws and pipe caps were checked to make sure none had shaken loose during the ride. By 1:00 P.M. everything seemed ready. Bill Bollay, whose paper had instigated all this commotion, was on hand to watch the young experimenters, as were two more Caltech students with cameras who had turned up to record the scene. As the rocketeers slumped on the riverbed, worn out from their exertions, one of the students took a photograph of the group.

It is an iconic image and shows the relaxed informality of proceedings. Schott sits at the far left of the picture, seemingly exhausted, staring into the camera. Apollo Smith, wearing his ventilated pith helmet, reclines on one arm. Malina stretches out next to him, looking slightly irritated. Was it because of something Forman, who sits next to him, has said? Forman is chewing a reed and reclining contentedly, as if he has just made a remark that rather pleased him. Parsons lounges at the right of the photo, wearing his trademark vest, his legs resting on a piece of scrap wood. He smirks into the dust in front of him. Whether he is laughing at Forman's joke or pleased that he is finally getting to test the rocket motor is hard to tell. Certainly he was always happiest when experimenting.

At the center of the photo stands the product of their long hours, in all its diminutive glory. The rocket motor was affixed to a spring, then attached to a three-foot-tall stand, positioned so that the exhaust flame would shoot from a nozzle straight up into the air. Forman had crafted the motor, less than a foot long, from gleaming duralumin, a lightweight but strong alloy of aluminium, copper, and magnesium, often used by the aircraft industry and most likely "borrowed" from GALCIT.

Connected to this motor were four hose lines. One supplied gaseous oxygen, another methyl alcohol under pressure, the third water for the cooling jacket that surrounded the motor. The fourth monitored pressure inside the chamber. Malina and Parsons would measure propulsion by the amount of thrust the motor gave to the spring: As the thrust increased, the spring mechanism would be depressed in direct proportion to the degree of thrust generated. The thrust was marked by a diamond-on-glass dial under the motor chamber. A pressure gauge told of the force *inside* the motor. If it got too high, the rocketeers could turn off the fuel to prevent the rocket motor from exploding. But for all its intricacies the rocket had no spark plug to light the mixture. The experiment was to be started with a simple cord fuse running out from the motor.

Anticipation ran high. The rocketeers connected and thoroughly tightened the fuel lines running to the motor. Check valves, flow meters, and pressure gauges were put in place. The rocketeers and their witnesses positioned themselves behind the sandbags. Parsons lit the fuse and rushed back behind the sandbags as Malina and Forman turned the fuel and oxygen on. The supply tubes snapped taut as the liquid and gas flowed into them. But the rush of air coming from the tubes blew the fuse out of the motor. Now methyl alcohol began to bubble over the still-cold rocket. The rocketeers rose slowly from behind their sandbags and shut off the fuel and oxygen valves. After recalibrating the equipment and emptying the motor, they crouched once more behind the sandbags. Parsons now tied the fuse in place and hurried back. But again the fuse was blown free and the methyl alcohol flooded unlit into the motor.

The group's frustration was palpable. Months of work had gone into the engine and now they couldn't even light it. It was as if they had built a car but forgotten to add a working ignition switch. The two student photographers decided that they weren't going to see any action and headed back to Pasadena.

Once more the rocketeers got off their hands and knees and wandered over to the rocket stand, now drenched in methyl alcohol. Parsons gave the equipment a perfunctory rub down and tied another fuse to the rocket. He lit it and ambled back to the sandbags, expecting the fuse to be blown loose from the chamber again. And it was, high into the air. However, in falling it ignited some of the alcohol that still remained on the equipment. A yellow-orange flame burst off the top of the nozzle. Parsons spun around and the rocketeers stuck their heads above the sandbags. Their equipment was not firing; it was on fire. Just as they started to move towards it to dampen down the flames, one of the rubber tubes still carrying oxygen to the chamber snapped free from its connection to the rocket motor and swung into the path of the flame, blasting a wall of fire directly towards the unprotected rocketeers. To a man they turned and ran across the valley floor as the flames flared behind them. Eventually the check valves that were attached to the oxygen tank clamped down, and the methyl alcohol remaining on the set burned off. Gingerly, the rocketeers returned to view the damage.

It was not a pretty sight. The rubber hoses were melted, and the brass couplings that held the oxygen hose to the rocket motor had snapped off. The cooling water jacket was leaking, as was the base of the rocket motor's nickel-steel nozzle, and the nitrogen regulator used to force the fuel into the motor had given up the ghost. Parsons and the rest of the group were, however, grinning from ear to ear. If only the photographers hadn't left, they lamented as they dismantled the setup; now that would have been a picture!

While they had not been able to fire the rocket, the flames had rekindled their excitement. After all, they argued, the fuel line had operated perfectly, not to mention the life-saving oxygen tank check valves, and they were beginning to understand the pressure at which the fuel and oxygen should enter the motor. As they dismantled the apparatus to take it back to the workshop, they persuaded themselves that whatever its failings, their experiment boded well for the next trial. As Parsons had shown before, rocket experimentation was akin to an addictive drug. Even less than triumphant attempts left one breathless to improve the rocket and try again. "As a whole," wrote an excited Malina to his parents the following day, "the test was successful."

The Rocket Research Group continued their experiments throughout the fall of 1936. On November 15, using a spark plug instead of a fuse, they managed to light the rocket for five seconds. But it burned with a high yellow flame, indicating incomplete combustion, and there was little noise. Seconds later the oxygen line caught fire, again clogging all the motor's parts with burnt rubber. By the end of November the rocket motor was repaired and ready for another test. New nozzles had been made and copper tubing replaced the rubber. This time the motor fired with a great noise for a full twenty seconds, straining against the stand. The rocketeers could hardly believe their eyes or ears. When the rocket finally sputtered to a halt, Parsons, Malina, Forman, and Smith crawled from behind their sandbags and cheered wildly.

On January 16, 1937, amidst an epidemic of flu at Caltech, the exhausted rocketeers made their fourth test. It was the best yet as the rocket motor fired for forty-four seconds straight. They prepared their results—listing the fuel and oxygen pressures, the varying pressures inside the motor, the height and color of the flame, the different thrusts gained with different nozzles—and placed them in front of Kármán. The Hungarian professor was impressed. Since they seemed to be serious about experimenting, he said he would allow them to conduct small-scale rocket motor tests on campus, rather than far away in the Arroyo. It was a decision he would come to regret.

As the rocket group rejoiced in their first success, Parsons received a visit from a man he had never met—his father, Marvel Parsons. In the immediate years after leaving Los Angeles, Marvel had lived a picaresque life. He had joined the army, becoming a champion shot, and had been part of the United States force that had chased the Mexican revolutionary Pancho Villa across Mexico. He had worked his way up the ranks to become a major in the coast artillery corps, a branch of the army responsible for setting up defensive forts at strategic points along the United

States coast. He had also remarried and had another son, Charles. Along with his family he had been sent to serve in the Philippines for two years. When he returned to the United States, Marvel felt he should reintroduce himself to his long lost son.

Up until this point the fourteen-year-old Charles Parsons had never been told of his father's other family. "[It] came as a total and complete surprise," he remembered. This visit was the only time that Charles would meet his half brother. Parsons showed Charles and their father his home laboratory and introduced them to Helen (though Marvel's former wife, Ruth, did not make an appearance). It was an awkward visit. "We didn't even stay for a meal," remembered Charles. Helen remembered how Jack seemed strangely unaffected by meeting his father. "He wasn't touched by the visit. I don't think he was touched by anything that wasn't the moon."

What Marvel thought about the meeting is unknown, for it was not long after this visit that he fell victim to a strange psychosis that would cloud the last years of his life and prevent any further reconciliation with his son. He suffered a heart attack and, because of a mistaken diagnosis, was given twenty-four hours to live. Marvel physically recovered within several weeks, but his mind was forever colored by what the doctors had told him. His date with death postponed, Marvel became fixated on his own mortality. His hospital record reports that he "persisted in the belief that his death was imminent...He showed considerable emotional instability, would cry very readily, had ideas of unworthiness, felt himself incapable of discharging his duties...He believed something was escaping from his chest...He suffered from marked mental depression and exhibited definite delusions and hallucinations." He attempted suicide and was committed to St. Elizabeth's Hospital in Washington, D.C., suffering from "melancholia." He would spend the rest of his life there, sharing the wards with 7,000 disturbed others—among them the poet Ezra Pound—in a permanent state of depression. He was not even given the satisfaction of an accurate prophecy.

He finally died from meningitis in 1947 and was buried in Arlington Cemetery.

Parsons would see his father once more when he traveled to Washington, D.C., on business in 1943. The experience would not be a pleasant one. "I have seen my father," he wrote to Helen; "he was sadly ill. It was a bit of a shock." He would not have appreciated the fact that the sole legacy left him by his father was the specter of mental illness lurking deep in the recesses of his imaginative mind.

Back at Caltech, the rocketeers were celebrating their coup. Their work could hardly be ignored by the scientific establishment once it was happening on campus. They would now try to build bigger and more powerful rocket motors; perhaps they might even impress someone enough to gain some funding, for the project was still being paid for entirely out of the rocketeers' own pockets. They were cutting back on all but necessities. Cigarette butts were being carefully hoarded and the tobacco rerolled; trips to classical music concerts were completely curtailed. But the privations did not seem to touch Parsons. As he would later write, his rocketry work was the only thing he was willing "to go without food and sleep, work like hell, throw money and cut throats for." When she could, Helen, who continued to work at her stepfather's business, would bail out the rocketeers with cash advances. At one point Parsons even persuaded her to pawn the diamond engagement ring he had given her for a few hundred dollars. Helen's finger would be intermittently lighter for years to come.

Once, when the ring was in the pawnshop and Parsons was still short of the money for a particular rocket part, Helen got her revenge. Refusing to lend him any more money until her house was cleaned from top to bottom, she forced Parsons and Forman to tie on pinafores and do the dusting and sweeping to earn their money. Helen made sure that every part of the house was spotless before she handed them the five dollars they needed for their rocket part.

When Helen had no more money to give, Parsons was forced back into extended stays at the powder companies. He would spend weeks at a time away from Pasadena, working twelve-hour shifts. In the evenings he would sit, with his pipe in hand—a new affectation—and write long melancholy letters back to Helen in Pasadena, telling her how "fed up + disgusted" he was with his life.

He began suggesting "more practical + remunerative ways to make money" and listed new businesses that the two of them should start up together: an exclusive lending library, a perfumery, a pipe shop, a restaurant, or a garden supplying florists. The list seemed to reflect Parsons' own favored inclinations rather than the practicality of such ventures. "Am kicking myself for a prize sap. I could have looked for any kind of job in a perfume—cosmetic or drug plant—or book store . . . If I had worked on that like the rocket we would probably be on top now, and I would be farther with the rocket."

Despite the complaints and his grim surroundings, there was still room for his usual lyricism. "I saw the sun set over the ocean. It was gold, + the clouds were black + fringed with gold, + black and gold rays streamed up across the sky; it was as though alchemistic doors were ajar briefly during some Tremendous Transmutation." And there were still words of gentleness to Helen. "Goodnight—dear—live as I do in the time when we will be together—work with me to make it soon."

By April 1937, the group had attracted a new member, a Chinese graduate student named Tsien Hsue-shen. Slight, short, and impeccably neat, Tsien arrived at Caltech in 1936. His remarkable aptitude for mathematics prompted Kármán to take him under his wing immediately, and Tsien swiftly became one of his star pupils. However, his aristocratic manner—he was descended from the tenth-century emperor Qian Liu Tsien—rankled the other graduate students. "He was [a] very stubborn, very individualistic fellow," remembered his roommate at Caltech, Shao-wen Yuan. "He always thought he was right, and usually he was. But he made a lot of enemies." This air of superiority gained the twenty-five-year-old Tsien the nickname "Son of Heaven."

Tsien shared an office on campus with Apollo Smith, who had himself grown tired of Tsien being "not talkative" and "really arrogant." Nevertheless, when Malina came to talk to Smith about rockets, Tsien could not help but be intrigued. Some of his haughtiness resurfaced when he found out that two untrained, non-Caltech members had initiated the project, but he managed to work with them as best he could. "His attitude to Parsons and Forman would have been ... 'they couldn't understand what he was doing anyway', but he didn't ostracize himself from them," remembered Malina. Although Tsien shied away from experimentation, he could be found working feverishly on the project in his study, making calculations and developing theories on the technical aspects of rocket flight, such as the air resistance on a rocket moving at high speed and the effects of exhaust nozzle angles on motor thrust. The group's research, their correspondence with the other rocket groups, and the results of their previous experiments were incorporated into a bulky, loose-leaf volume which they called the "Bible." They were beginning to build a sizeable background of original material to work with, even if the operation was still destitute.

There are times in the struggles of scientific pioneers when something akin to divine providence plays a hand. In late April, Malina gave a talk on the rocketeers' findings to Caltech's weekly seminar program for engineers and science students. He thought nothing more of it until two weeks later when he was approached by Weld Arnold, an assistant from Caltech's astrophysical laboratory. Arnold had been fascinated by the talk, he said. Then he coolly offered Malina the unbelievable sum of

$1,000 to aid the rocketeers in the continuation of their work. There was one condition: He could join the project as official photographer.

Malina and the others were amazed, not only by the size of the donation but by its source. None of the rocketeers seem to have known Weld Arnold in the slightest. He was, after all, twenty years older than them. What's more, Arnold was hardly a typical benefactor. He rode his bicycle the five miles from his home in Glendale to Caltech every day, and his job as a laboratory assistant was hardly well paying. That did not keep the rocketeers from accepting the gift with alacrity. Mindful of endangering this mysterious windfall, they decided not to ask questions about its provenance. And Arnold did not even hint at the money's source. A week after his proposition, he returned to Malina with the first five hundred dollars in a bundle of one- and five-dollar bills wrapped in newspaper. The other five hundred, he said, would follow shortly.

Malina immediately took the money to Clark Millikan, nonchalantly placed it on his desk, and asked how he could open up a fund with the institute for the use of rocket research. Millikan, who had spurned their project a year before, was "flabbergasted." The money became known as the Weld Arnold Rocket Research Fund. The Rocket Research Group was now officially recognized by GALCIT not only as an affiliated project, but as a fully financed one as well. Its first members were Parsons, Malina, Forman, Tsien, Smith, and Arnold, their mysterious benefactor.

Their newfound solvency, however, did not do much to legitimize them in the eyes of many of their scientific peers. When Malina went to speak to Dr. Fritz Zwicky, a professor of physics and one of the most feared teachers on campus, about some of the theoretical problems their rocket research had posed, Zwicky erupted. Malina later recalled, "He told me I was a bloody fool, that I was trying to do something that was impossible, because rockets couldn't work in space." It was the same

erroneous argument that the *New York Times* had used to discredit Goddard back in the 1920s. Malina was chastened by the
virulence of the attack, but not put off. Every other university in
America still ignored rocketry as a legitimate science. Although
the rocketeers had been brought into the fold, they were by no
means part of the established scientific flock.

Arnold's beneficence allowed the group more time in between paid work to concentrate purely on constructing a new
rocket motor and test apparatus. "The work is great fun," wrote
Malina to his parents; "we are looking forward to three weeks
of head scratching." They now planned to test various oxidizer-
fuel combinations inside the venerable GALCIT building itself.
The suspicious university authorities had been somewhat molli-
fied by the rocketeers' insistence that they were only going to
use a miniature motor to run the tests: Crafted by Forman, it
had a nozzle less than an inch in diameter. Little did the powers-
that-be know that the rocketeers had an uncanny ability to cause
trouble, no matter the size of the rocket.

On a Saturday morning in August, when the GALCIT building was almost deserted, Parsons and Forman met in front of its
imposing copper doors. Malina and Smith were off siphoning
nitrogen tetroxide, to use as an oxidizer, from a cylinder that
had been left for them on the lawn in front of the chemistry
building. After extracting about a liter of the highly corrosive
liquid, they tried to shut off the valve but found it jammed.
Smith and Malina frantically tried to stop the fuming fluid from
gushing out, but, overcome by toxic vapors, they were finally
forced to retreat to a safe distance. They watched as the cylinder emptied its entire contents onto the prim green lawn, the
grass wilting and turning brown in front of their eyes.

While Malina and Smith were discoloring Caltech's prized
turf, Parsons and Forman had been erecting an extraordinary
apparatus within the GALCIT building. It consisted of a fifty-
foot pendulum which hung down the side of a narrow open staircase, from the third floor down to the basement. At the bottom

of the pendulum, they bolted the tiny rocket motor, attached by tubes to the tanks of nitrogen tetroxide and methyl alcohol. When the motor fired, its thrust could be measured by gauging the distance the pendulum swung. The apparatus and procedure were an ingenious idea. However, almost as soon as the rocket was lit, it misfired—a seal broke, or a fuel line sprang free—and released a cloud of highly toxic, highly pressurized nitrogen tetroxide and alcohol. Once again the choking and spluttering rocketeers were forced to flee, and it was agreed that experiments should be halted for the day.

What the rocketeers didn't realize was that the corrosive fog they had unleashed had permeated most of the GALCIT building. When they reconvened the next day, having carefully avoided the irate gardeners congregated outside the chemistry building, they were met by GALCIT's furious janitorial staff. The caustic gas had left a layer of rust on virtually all the metal scientific equipment in the building. The rocketeers were forced to spend the rest of the day scrubbing it off with oily rags under the baleful stare of the faculty.

Kármán, hearing about his staff's antagonism towards the motley crew of researchers—some of whom, it was pointed out, were not even students of the institute—expelled them outside to a concrete platform attached to the corner of the building. Here they suspended their pendulum from beams that stuck out from the roof and built a small laboratory from sandbags and corrugated iron.

Having escaped explosions in the Arroyo and asphyxiation in the GALCIT building, the Rocket Research Group swiftly became known among the other students as the "Suicide Squad." Parsons cherished the peculiar personal chemistry of the group and the informality with which it worked. It was a fluid, self-sufficient body, the members responsible only to themselves. Within the group the friendly friction between the trained scientific mind and the untrained imagination caused ideas to crack and sparkle. Experimental results led to theory, and theorizing led back to experiments. The group knew that trial and error was the only way to progress in rocketry. If members of the Suicide Squad wanted to walk away in a sulk, the others knew they would eventually return. After all, where else could they get to play with rockets all day?

By placing the Suicide Squad out in the open, Kármán might have saved his laboratory from destruction, but he also ensured that the rocketeers gained notoriety. Their continuing tests—some comical, some successful—attracted much attention on campus. Loud intermittent pops and blasts would emanate from their laboratory. "I remember we ran to the window when we heard one bang and [saw] one of the guys . . . thrown out of the hut," remembered aeronautics professor Hans Liepmann. "[They] were a strange bunch, a very strange bunch." Between classes other students would crowd around the group to watch them prepare their experiments. They became such a campus sideshow that one of the Suicide Squad suggested a collection be taken from the gawking students to help finance their work. Many of the Caltech students hoped for a spectacular explosion, but the Suicide Squad were taking no chances. "About every part of our apparatus is a research problem in itself," wrote Malina to his parents. "Everything is being made 5 or 6 times stronger than necessary. The $1000 fund is rapidly diminishing."

The rocketeers were also achieving notoriety off campus. Newspaper reporters from the local Pasadena papers began writing articles on the group. Intrigued by the rocketeers' affiliation with Caltech, *Popular Mechanics* ran a photo of their experiments. Nevertheless, within the Caltech hierarchy they were still treated with caution and some amusement. At a GALCIT seminar Malina was asked to review a paper on the equally embryonic—but quite unrelated—technology of helicopters. "I seem to be becoming the department fantasy expert," he lamented.

While the group was egalitarian to a fault, outsiders to the project often overlooked Parsons and Forman's involvement. The

California Tech, Caltech's student newspaper, carried a front-page article on the rocket experiments that mentioned only graduate students Malina, Smith, and Tsien. Shortly afterward, a small article appeared in *Time* magazine, but once again it was only the Caltech students who were named. To both Parsons and Forman, their exclusion must have come as something of a blow, especially to Parsons who had longed for a university education. Although he had been denied a place at Stanford because of his precarious financial situation, Parsons had eventually been admitted to the University of Southern California in Los Angeles to take night courses in chemistry. He had been attending for just over a year, but his overwhelming workload at the powder companies and on the rocket project meant his presence was sporadic. Now, just when it seemed he was being admitted into the Caltech fold, he found himself still being kept at arm's length. But he was soon to find his skills acknowledged in a slightly more unexpected venue.

While Pasadena basked in its aura of refinement, a few miles away in Los Angeles, a story straight out of a hard-boiled Raymond Chandler novel was unraveling. Los Angeles had been hit hard by the depression. It suffered the highest bankruptcy rate in the country and unemployment was rife. By 1933, for all its Hollywood trappings of glamour, the city had become one of the most depressed in America. Locals went broke, impoverished transients arrived, the homeless slept on the street. Soup kitchens abounded, offering a bowl of brown rice and vegetable soup for a penny. With poverty came crime. Under the crooked smile of Mayor Frank Shaw, Los Angeles had been skillfully worked into a ferment of vice. Since Shaw had been elected in 1933, as a Roosevelt-idolizing Democrat, both panhandlers and corrupt executives had flourished. Prostitution prospered, gambling was endemic, and bribery was a way of life. The cops were on the take, and Police Chief James Davis openly boasted of his "shoot-first-ask-questions-later" policy. Even the district attorney was operating under a felony indictment. By 1937 an estimated six hundred brothels, three hundred gambling houses, eighteen hundred illegal bookies, and twenty three thousand prohibited slot machines were scattered across the city.

It was left for privately funded reform groups such as CIVIC (Citizens Independent Vice Investigating Committee), an organization founded by the reformist cafeteria owner, Clifford Clinton, to challenge this corruption. CIVIC looked into rumors of corruption in local government, most noticeably the appointment of Mayor Shaw's brother, Joe Shaw, as his "private secretary," a cover for his real role as the chief fixer in City Hall. In 1937 Clinton filed a county grand jury report charging that the local government was hopelessly corrupt and identifying more than a thousand gambling and prostitution rackets allegedly under the protection of Shaw's administration. For this, Clinton faced a barrage of abuse from all those who had a finger in the Shaw pie, most noticeably the *Los Angeles Times,* which labeled Clinton, not Shaw, "Public Enemy No. 1." (Shaw had personally engineered a vastly inflated sale of the paper's former property.)

In retaliation for his report, Clinton's house was bombed. His wife and children narrowly escaped with their lives. When the police were asked to investigate, they claimed that Clinton had engineered the explosion as a publicity stunt. His cafeterias were stink bombed, and were besieged by health regulation officials claiming unsanitary conditions. Hundreds of nuisance suits were filed against him from "patrons" claiming that they had been food poisoned. But it wasn't just Clinton who found himself under attack. On January 14, 1938, Harry Raymond, a former Los Angeles Police Department (LAPD) officer now operating as a private detective for Clinton's CIVIC organization, pressed the starter button of his car and ignited a bomb placed underneath it. The car was destroyed, the garage lay in tatters, but somehow Raymond—despite bleeding from more than fifty head and body wounds—survived. Even this jaded city was shocked when evidence pointed to none other than Police Captain Earl Kynette, the head of police intelligence, as the culprit.

Kynette was arrested along with two of his detectives and charged with attempted murder.

The trial began on April 12 and dominated news up and down the West Coast. The prosecution immediately asserted that Raymond's car bombing could lead all the way back to Mayor Shaw himself. Over the next three weeks, as witnesses were called, the special prosecutor started receiving bomb threats, and the police chemist had to be given an armed guard. Tension was running high by May 9, when the prosecution called on its youngest witness to date: John W. Parsons.

A vital piece of evidence was a stretch of fuse wire, the same as that used in the car bomb, found in Kynette's garage. Now the prosecutors had to show what kind of a bomb could be constructed using this fuse wire and whether that bomb could have been used in the Raymond bombing. When they approached Caltech, they were informed that if knowledge of explosives was needed, then the twenty-three-year-old Parsons was their best bet. That Parsons was held in such high regard by Caltech's officials, despite both his age and his loose links to the university, attests to his extraordinary knowledge of chemistry.

As he had progressed from blowing up the lavatories at the military academy to shooting off rockets, Parsons had learned how to control explosions, to temper their bite, and to recognize their character. To him an explosion was not a violent and meaningless release of energy but a dance of expansion, one that could be studied as one might study a symphony. He also knew which chemicals were hard to get, which could be stored safely, which were too powerful, which were not powerful enough. He read not just the mangled car and tattered wall but the exploded casing of the bomb itself, as clearly as a book. Indeed, it was the smallness of these fragments, collected at the blast site, which gave him his most important clue.

Having seen so many of his own rockets explode and having had to pick up the pieces afterwards, Parsons knew all too well what different fuels did to different casings. The bomb container in the Kynette case had been torn into tiny pieces. After making numerous tests he narrowed the possible explosive down to a high explosive, a nitrocellulose-based, smokeless powder similar to those he had helped create at the Halifax and Hercules powder companies. In fact, Parsons had even considered using it as a fuel for his own rockets.

Parsons built a bomb that he predicted closely resembled the one used in the Raymond car bombing. Then, accompanied by the court's special prosecutor, he blew up an old Chrysler with it. The results of the explosion, specifically the size of the bomb casing fragments, were almost identical to those found at the Raymond site. His evidence was good enough for the prosecution: He would stand witness.

His appearance in court was accompanied by a flurry of articles and pictures. He may have been one of a number of scientists called to the stand that day, but only he was young and charismatic. Most importantly from a newspapers' point of view, he blew things up. EXPLOSIVES EXPERT MAKES BOMB REPLICA, read the Los Angeles Times; CALTECH MAN TELLS OF BOMBS, blared the Pasadena Star-News.

Parsons brought a duplicate of the bomb to the stand. Six inches in length and three inches in diameter, it was made of cast iron water pipe; he had noticed that the original bomb's fragments had born the marks of "threads"—the grooves made for screwing pipes together. The duplicate was a persuasive piece of forensic science and a fair bit of courtroom theater. The "appearance of a very business-like looking bomb in the hands of a State's witness late yesterday caused turmoil at the trial" read the Los Angeles Times. The bomb was so realistic, Parsons' description of it so menacing, that the presiding judge was uncertain whether the replica contained explosives or not. He refused to allow the jury to touch it out of safety concerns. "It was a very colorful piece of evidence," sniffed the Times. It was also an example of how persuasive and compelling Parsons could be.

Perhaps thinking that Parsons' youth could be a chink in the prosecution's armor, the Kynette defense team began a fierce cross-examination of him. If they could discredit the "young expert," the prosecution would be severely hampered. But, again, Parsons performed well in the spotlight. He even gained a laugh out of the packed spectators with his unflustered delivery in the face of an increasingly frustrated defense. When Parsons was asked what the explosive velocity of trinitrotoluene was, the court stenographer was flummoxed by the spelling of the word. Parsons coolly suggested that its more common abbreviation, "TNT," might make things easier. The gallery, having sat through hours of technical jargon, erupted in laughter.

Parsons left the stand with his head held high, his standing as an "expert" confirmed. On June 16, 1938, Kynette and his two assistants were convicted and sent to San Quentin. So damning was this case in the eyes of voting Angelenos that soon afterwards Shaw became the first big-city mayor in United States history to be thrown out of office by recall. For Parsons the case was a personal triumph. Not only did it highlight his prodigious chemical knowledge, but it also prompted his recognition as a bona fide member of the scientific community. The *California Tech* had failed to mention him in its article on rocket research, but during the Kynette trial he had appeared in the pages of all the major West Coast newspapers as not only a scientific expert, but a Caltech man.

5

. . . .

FRATERNITY

*Civilization has absolutely no need of nobility or heroism.
These things are symptoms of political inefficiency. In a
properly organized society like ours, nobody has any
opportunities for being noble or heroic. Conditions have got
to be thoroughly unstable before the occasion can arise.*

—ALDOUS HUXLEY, *Brave New World*

Among stories of riots in the campus dormitories, reports on
fraternity banquets, and the occasional mention of rocketry
experiments, global events were rapidly filling the pages of the
California Tech. FASCISM OR COMMUNISM, WHICH MOST TO
FEAR? ran a headline from 1937. It was a timely question:
Events in Spain showed the two ideologies locked in direct op-
position as General Franco's Fascist-backed forces battled the
Second Spanish Republic, supported largely by anarchist and
Communist groups.

The effects of both political movements could also be felt
closer to home. The rise of Fascism in Germany and Italy had
prompted European Jews to emigrate in a steady flow to Los
Angeles and Pasadena. Some who had been banned from teaching in Nazi Germany took posts at Caltech. The Suicide Squad's
mentor, Theodore von Kármán, had himself suffered similar maltreatment years before while teaching at Aachen. He

now grasped every opportunity to question those refugees who made it to Pasadena, and with increasing incredulity he repeated to his students, "The stories are all the same! The Nazis are really devils!"

Communism, thanks in part to the very virulence of its opposition to Fascism, seemed a more attractive and intellectual option, especially to students. The successes of Franklin D. Roosevelt's socialistic New Deal program, with its support of the Communist-led unions, had undoubtedly leant much authority to leftist thought. While some, like Huey Long, the famed Louisiana senator, decried the New Deal as containing "every fault of socialism...worse than anything proposed under the Soviets," among the intellectual community Roosevelt's program was roundly praised. The German writer Thomas Mann, now living in Los Angeles, declared Roosevelt "an American Hermes, a brilliant messenger of shrewdness." For some of the students at Caltech, an alternative to the capitalism that had caused the depression and affected so many of them directly must have been of real interest.

The turbulent politics of the time was reflected in Parsons' latest plan to gain funding for his rocket research. Every Monday he and Malina would meet and work on a novel which they hoped to sell to one of the nearby film companies. Its subject was the fantastical world of rocketry. All that survives of the book is a chapter-by-chapter summary, no doubt intended as a proposal for publishers or Hollywood agents. It initially appears as a histrionic roman à clef offering a skewed take on the members of the Suicide Squad. The story is also remarkably prescient, blithely foretelling the postwar repercussions of the rocketeers' political beliefs. Yet nowhere does the story come closer to prophecy than in relation to Parsons himself, prefiguring not only his eventual absorption in the occult world, but even the explosive manner of his death.

The untitled novel is set in "the Institute," a dead ringer for Caltech. The hero is Franklin Hamilton, a genius physicist and rocket scientist. With his "black, closely cut slightly curly hair" and "good looking, sharply chiseled face," he seems to be a replica of Parsons himself. Furthermore, Hamilton is "not the type usually thought to be scholastic" and his colleagues refer to him as a "lady's man." Like Parsons he constantly smokes cigarettes, placing the butts in his waistcoat pocket.

Other members of his rocket group include Lin Lao, who is faced with the dilemma of going back to China or staying and working on the rocket project; Thomas Elwood, a Nazi-hating union organizer; and a recently defrocked Franciscan monk named Theophile Belvedere, who has an overpowering interest in the Cabbala. All these characters share traits with the members of the Suicide Squad. Lin Lao is clearly a study of Tsien, who had been thinking of returning to China as the Sino-Japanese war of 1937 escalated; Elwood with his social conscience is pure Malina; and Belvedere's interest in mysticism is a product of Parsons' own embryonic interests in the esoteric world.

This hodgepodge of union organizers, genius scientists, and persecuted minorities engage in "preliminary experiments to learn the type of problems they must solve to construct a successful sounding rocket," exactly as the Suicide Squad had planned. Their greatest trouble, as in real life, is a lack of financial backing. A test is made, with a setup much like the one used in the Arroyo Seco, but as the second rocket motor prepares to fire, Belvedere inexplicably attacks the apparatus with a wrench, striking the fuel tank and "causing a terrific explosion." Once the smoke has cleared, the group finds him badly injured but conscious enough to reveal the tragic love affair that forced him to leave his monastery. He dies from his wounds in a manner eerily similar to Parsons' own death by explosion. Part thriller, part Brecht, the story dashes headlong into espionage, murder, and organized labor. Franklin Hamilton is roundly attacked for allowing Elwood, a pro-union organizer with "un-American beliefs," to be connected with the institute. He is saved from being forced to resign by a sudden donation of $100,000 by Franz

Adams, a wealthy aircraft manufacturer, union hater, and Fascist-sympathizer, who has decided to sell the rocketeers' plans to the Nazis. In the denouement the "Institute" is pilloried for housing un-American scientists—another event that would actually come to pass—and the rocket plans are about to leave on a plane for Nazi Germany. Fighting to save the day, one of the rocket team wrestles with the Nazi messenger, knocks him out, and grabs back the blueprints. When he receives them, Hamilton must face the grim realization that for mankind to be truly safe he must destroy all his research about rockets. "Only harm to humanity can at present result from its knowledge," he intones. The plans are burned, and the rocket prototype they constructed is fired into the air, to be lost forever.

Parsons and Malina sent the summary to MGM studios, but perhaps unsurprisingly, nothing appears to have come of it. Still, the outline of the novel proved to be a strangely unsettling and prophetic piece of work. Not only did it predict so many features of the rocketeers' future lives, but it also foresaw the Nazi's secret obsession with rockets, barely hinted at until now.

The informal discussion groups that Parsons, Malina, and the other members of the Suicide Squad held at each other's houses had taken a new direction ever since Malina had invited a research assistant in the chemistry department at Caltech, Sidney Weinbaum. Born in 1898 to a middle-class Jewish family in the Ukraine, he fled his homeland after the Russian Revolution and arrived at Caltech in 1922. With an aquiline nose, pointed ears, and humped shoulders, he was unattractive, but those who knew him recognized that he was something of a Renaissance man. He had studied medicine at Warsaw University and physics at Caltech; had won the Los Angeles chess championship twice; and was an accomplished concert pianist. Older and worldlier than the young rocketeers, he could, it seemed, turn his mind to anything.

Weinbaum brought with him a motivating political presence. Until now the rocketeers' talk had focused on poetry, politics, music, and movies. Weinbaum bent their conversations increasingly towards politics. They needed little encouragement. The group had already talked much of its socialist sympathies and its hatred of Fascism, and Weinbaum, a member of the increasingly influential American Communist Party, was not ashamed to display his credentials. He had already circulated a petition through Caltech calling for the international recognition of the Soviet Union, and he was known for his pro-Soviet harangues. In a small and relatively conservative society like Caltech, such a stand could make others uncomfortable. Aeronautics professor Hans Liepmann thought Weinbaum was "a nut ... He never made any beans that he was a communist." Kármán dealt with him in his usual mischievous manner, introducing Weinbaum at one of his parties by saying, "Here comes my friend Weinbaum, he deals in chemistry and communism." Weinbaum soon convinced Parsons, Malina, and Tsien (Forman and Smith seemed not to have accepted Weinbaum's offer) to stop their own meetings and come together at his bungalow in Pasadena with others from Caltech. His motive was simple: Both Weinbaum and Frank Oppenheimer, brother of the famed physicist Robert Oppenheimer and a graduate student at Caltech, had been asked by the American Communist Party to try to organize a Communist group at Caltech.

To begin with, the meetings were not especially militant, but they were intoxicating. "It seemed a mental exercise for these men," remembered Liljan Darcourt, Frank Malina's fiancée at the time. "They discussed Russia [and] Trotsky ... If you had half a brain you were interested in it. There was a great deal of misery at the time, the Depression and a president who couldn't do a damn thing ... They were all enormously patriotic, they felt for the people." Soon Parsons subscribed to the *People's Daily World,* the first openly Communist newspaper in the country, and became a member of the American Civil Liberties Union, which at the time was labeled a "Red" organization.

The meetings were usually scheduled for Wednesday nights. Between twenty and thirty students, staff, and a few non-Caltech townspeople congregated at Weinbaum's house. The meetings followed a loose format, as they had always done, but the discussions of the Spanish Civil War were marked by their increasing passion. Once voices had dropped and tempers cooled, the usual mixture of music and parlor games took over. Tsien would sit in the corner playing his recorder, while the others gamely tried to beat Weinbaum at kriegspiel, a variant of chess. Meeting after meeting, Weinbaum continued to push the group further into politics. He was soon providing the students with books and pamphlets, offering to loan or sell them copies of Lenin's *State and Revolution,* Marx's *Capital* and *The Communist Manifesto,* and Stalin's *Messages.* It was not long before the rocketeers' informal discussion group had been transformed into Professional Unit 122 of the Pasadena Communist Party and Weinbaum started pressing attendees to become card-carrying members of the party themselves.

It was not illegal to be a member of the Communist Party at the time, but it was risqué and likely to be frowned upon by the college authorities. "It was essentially a secret group," remembered Frank Oppenheimer. "[Most people] were scared of losing their jobs." Even Helen Parsons was kept away from the meetings. "I would deliver Jack ... a block or two from where the meetings were held, so that I wouldn't know, and nobody else would know [where the meeting took place]." As the meetings grew more political, Weinbaum suggested that it would be to Parsons' advantage to become an official member of the Communist Party.

Obliged to make a choice, Parsons chose not to join. He had always been what Malina described as a "political romantic," more anti-authoritarian than anticapitalist. When questioned by the FBI later in his life, Parsons said he had disapproved of

Weinbaum's Communist proselytizing. His denial seems slightly disingenuous, however. His interests in left-wing causes had always been strong, and he must have always known that Weinbaum was a Communist. More likely Parsons was disappointed that the group had been driven from its freewheeling discussions of politics, art, music, and literature towards the biting realism of Marxism in action.

His refusal to sign up cooled relations between him and Weinbaum. Soon he stopped attending the meetings altogether. Malina joined the party under a pseudonym, as did Tsien, a choice they would both regret when the postwar Communist witch-hunts began. That Parsons did not enroll caused no antagonism among the rocketeers. "We were a peculiar, closely knit group with very many cross-interests which threw us together," recalled Malina, "but the unifying thing with our group was rockets, you see."

Walking into Clifton's Cafeteria on South Broadway in downtown Los Angeles was—and still is—a bizarre experience. Founded in 1931, the multilevel canteen was decorated like a redwood glade, with a twenty-foot-high waterfall cascading into a stream that meandered silently through the dining room. Painted plaster redwood trees disguised steel columns; a forest mural covered the whole of one wall; and artificial shrubs and bushes were everywhere. An organist played softly, and high among the artificial rocks stood a tiny chapel complete with neon cross. However odd the décor, under the guidance of crusading political reformist Clifford Clinton, this kitsch cafeteria provided millions of low-priced meals to the out-of-work and destitute during the darkest days of the depression. Clifton's provided a surreal sanctuary from the broken world. Now, in 1938, the only meals it provided free were those that teenagers "chiseled" while no one was looking. Many of these boys would then sneak their spoils upstairs to a small room where around twenty to thirty people met on the first and third Thursdays of each month. Their ages varied from fifteen to forty years, although the average age was around twenty. Like the meetings at Sidney Weinbaum's house, these gatherings were not illegal, but membership was frowned upon. And, similarly, the topics for discussion involved the future—for these were gatherings of Los Angeles Chapter Number 4 of the Science Fiction League (LASFL).

Hugo Gernsback, the great proponent of science fiction and founding editor of the pulps *Amazing Stories* and *Science Wonder Stories,* had launched the Science Fiction League in 1934. It was intended as a nationwide club; local groups of science fiction fans would apply for a charter to found their own chapters. The League flourished and clubs were set up across the United States, the United Kingdom, and Australia.

Completely apolitical and hermetically sealed from everything in the outside world but discussion of the future, LASFL consisted of fans of the genre and those interested in cutting edge science. Some of the younger members appeared at meetings smoking pipes like their scientist heroes and clutching large scrapbooks filled with clippings on the march of technology. For most fans science was the doorway to the futures promised in the pulps. "It was still possible to regard science and science fiction as in some way parallel," remembered the science fiction author and occasional LASFL guest, Jack Williamson, "both engaged in the exploration of the frontiers of the possible, science fiction with the imagination instead of the telescope."

The members' scientific interests were legion, but foremost among them was rocketry. Like Parsons and Malina, the members of LASFL had an absolute faith that rockets would soon be built to allow man to explore the infinite reaches of space. Many were members of the American Rocket Society or similar local rocket groups, and some had even corresponded, like Parsons, with Wernher von Braun in Germany, about both rockets and science fiction. (Braun was such a science fiction fan that during the coming war he would maintain his much prized subscription to the science fiction pulp *Astounding Science Fiction* through a mail drop in neutral Sweden). Because of these space dreams and the precocious zeal of its fans, science fiction often received the same ridicule that rocketry had suffered. Jack Williamson recalled that when he had undergone psychoanalysis in the 1930s, he had been warned that writing science fiction was symptomatic of profound neurosis.

Just as Professional Unit 122 had been drawn together by the charisma of Sidney Weinbaum, so LASFL was run by an equally fascinating figure. Forrest Ackerman had happened across the first issue of *Amazing Stories* at the age of nine, and he had been reading and collecting science fiction ever since. Born in 1916 he had lived most of his life in Los Angeles. He had spent a year at the University of California at Berkeley, but disappointed at the lack of journalism courses there, he had returned to Los Angeles to look for work. Now twenty-two years old, he worked for the Academy of Motion Picture Arts and Sciences, editing the *Players Directory*, a casting catalogue of actors and extras provided for the Hollywood film industry. His knowledge of film was unparalleled and he was often asked to introduce features at the academy's theater. He took advantage of these invitations to ingratiate himself into the confidence of many of the actors and directors who came to watch.

As the self-proclaimed Number One Fan of science fiction, Ackerman felt obliged to know more about science fiction than anyone else. He was famed for watching 365 films in one year, and for corresponding in 115 simultaneous letter exchanges with other science fiction fans throughout the world. He edited fanzines, was fluent in Esperanto, and had created a new language called "Ackermanese," consisting mainly of abbreviated words and puns. His interests were so diverse that he would soon start a business called "Assorted Services," which offered to do anything for anyone, from walking the dog, through gardening, to returning library books. However, at the center of his frenetic life was one constant—science fiction, or as he described it, "my god and my religion."

With his many correspondences, Ackerman had his finger on the rapid pulse of the science fiction world. He attracted many seminal speakers to the chapter, most of whom were delighted to talk to an audience enchanted by their every word. Eric Temple Bell, professor of mathematics at Caltech, tried out on the group his latest stories of rapid and uncontrolled evolution. The astronomer Robert S. Richardson lectured to a rapt audience on his specialty, Mars, the muse of so many science fiction stories. (He would later write science fiction under the pseudonym Philip Latham.) The psychiatrist David Keller, a world authority on shell shock, read his grimly pessimistic tales of the end of civilization. LASFL members who worked at the aviation companies surrounding Los Angeles managed to persuade engineers and technicians to speak about the latest developments in flight technology.

Full-time pulp writers also came to give talks. E. E. "Doc" Smith (who had gained his learned nickname through working as a food chemist specializing in doughnut mixes) spoke about his famed *Skylark* "space opera" series, and a young L. Ron Hubbard—years from founding the religion of Scientology but already a pulp writer of some note—spun tall tales about how he had lassoed polar bears in Alaska and gave awe-inspiring examples of his skills at hypnotism.

Following an Ackerman-organized LASFL field trip to Caltech, during which the young members must have stopped by the GALCIT test shack, Parsons was enticed to speak at the meeting held on May 1, 1938. He cut quite a figure when he arrived, immaculate as ever in suit and tie. "My impression of him was like a young Howard Hughes," remembered Ackerman. He talked about the latest developments in rocketry at Caltech, about his group's successes and failures, and about the future and the moon. He roused his audience, especially one of its junior members, a young Ray Bradbury.

Bradbury was an eighteen-year-old newspaper boy at the time, still living with his parents in Los Angeles. He had a great interest in science fiction and had found out about LASFL through the letters page of the pulps—the great forum of debate for science fiction fans. He had always wanted to be a fantasy writer, and LASFL provided him with a community which shared his enthusiasms and understood his ambitions. When Parsons finished speaking, Bradbury asked him a barrage of questions. Many years later Bradbury remembered the incident well. "A young man, some six or seven years older than myself, was there, talking about Rockets and the Future," he recalled. "He was wonderful. I chatted with him, after, but was afraid of him because I was an uneducated non-student, a newsboy... and with no way of joining the rocket society the gentleman spoke of." Little did Bradbury realize how lacking in credentials Parsons was himself.

Though he eschewed membership in LASFL just as he had in the Communist Party, Parsons visited the group on occasion. He even invited a few of the members out into the desert to watch him and Forman set off some of their homemade black powder rockets. It must have been heartening for Parsons to be among real enthusiasts. Soon, however, the Suicide Squad's fan base was to get significantly larger.

On July 14, 1938, the millionaire sportsman and heartthrob Howard Hughes, accompanied by four copilots, landed in New York. He had traveled around the world in three days, nineteen hours, fourteen minutes, and ten seconds. It was the fastest circumnavigation of the globe to date, and 30,000 people showed up to cheer him home. Hughes, gaunt and weary to the point of exhaustion, declared, "All I can say is that this crowd is more frightening than anything else that has happened the past three days." Papers bragged it was the "greatest flight of all time."

Sharing the West Coast newspapers with Hughes on the day of his return were the prophets of a new flight technology—Parsons and Forman. In the *Los Angeles Examiner,* under the headline SAVANTS TEST "MACHINE GUN" MODEL, the exquisitely well-groomed Parsons and Forman were pictured firing a black powder rocket of their own construction. PASADENA MEN AIM AT ROCKET ALTITUDE MARK, read the *Pasadena Post*. They had been working on an impulse rocket, which could, as Parsons told the newspaper, "reload and fire between 100 and 200 gunpowder cartridges [when it was airborne] at the rate of ten explosions per second." He went on, "We plan to ... [fire] the rocket from a ground mortar to give it initial ground clearance before the rocket motor starts functioning." In essence the model was very similar to that described by Cyrano de Bergerac in his tale "L'Autre Monde," written almost three hundred years before.

The newspapers, who had always found that rockets made good copy regardless of the accuracy of their reports—witness Robert Goddard's suffering at their hands—had begun to take to the Caltech group. In June 1938, Malina had had lunch with the chief science writer for the Associated Press, who was making a cross-country trip "looking for the spectacular." Now reporters came daily to the GALCIT test shed, trying to get the rocketeers' prediction for the date of a moon landing. Throughout that summer articles appeared in the Pasadena and Los Angeles papers and were in turn reprinted by national technical magazines like *Popular Science*. The more conservative Malina was abashed at the sensational interpretations he read, complaining that "the reporters seem to have better imaginations than we do." Even a radio station approached the group, wanting to record the sound of one of their rockets for broadcast.

In response to all the press, crank letters flooded in. Some people offered their services as astronauts on the first rocket into space, while others offered secret designs that would help the Suicide Squad solve all their problems—for just 5 percent of future sales. One letter came from a circus daredevil. "Since about 40 years I am in the Sports and Show Business, and understanding I will have to retire soon, I will be very happy to honor one last time, my old name and reputation, giving to the crowd, the biggest Dare Devil Acrobatic sensation of all times: 'The Human Rocket'." Some persistent teenagers were actually granted an audience with the rocketeers. After all, it was not so long ago that they themselves had been teenagers and had had similar dreams. A few of the adolescents were kept on as helpers throughout the summer holidays.

The event perhaps most important in changing the professional perception of rocketry in the United States was Malina's appearance at the Institute of Aeronautical Sciences (IAS) in New York. The IAS had been founded in 1932 out of the rising success of aviation. In was intended as a forum for the aeronautical sciences and technologies, and Orville Wright was its first honorary fellow and Theodore von Kármán one of its founding fathers. Now at Kármán's instigation Malina traveled to the East Coast to give the first paper ever presented to the institute on rocket flight. The presentation was a significant step toward the rocketeers' recognition by their scientific peers.

Still, industry and the government were not convinced to take the rocketeers seriously. Their only financial support was Weld Arnold's mysterious one thousand dollars. They had hoped that Kármán's influence in Washington, D.C.—he had strong links with the army air corps—would help them gain official (and certain) rather than providential funding. But when Kármán invited a "big shot" from the army ordnance to visit Caltech, the big shot declared that rockets, as far as he was concerned, stood no chance of being used for any military purpose.

Malina, for one, was relieved by this. "My enthusiasm vanishes when I am forced to develop better munitions," he wrote. Parsons, however, was dreadfully disappointed. Although he was hardly a rampant militarist, rocketry was still his one and only dream. While others could fall back on their Caltech degrees as a safety net, Parsons and Forman were working for nothing but their own enthusiasm. The Weld Arnold fund could not last forever, and a military contract, while morally questionable, would at least allow Parsons to follow through on the years he had dedicated to rocketry so far. It would also assure him and Helen of a regular income.

Parsons' work on the "machine gun" type of rocket was stalling. If ten explosions were to happen a second, literally punching the rocket through the air, an incredibly fast reloading mechanism was needed, and Parsons and Forman had found no practical solution for making one. "Parsons is spending most of his time on the powder rocket," wrote a resigned Malina to his parents. "He is beginning to run short of money, so may have to take a job for a while again."

A general malaise hung over the Suicide Squad's research work through the autumn. On September 26, 1938, Frank Malina and his fiancée Liljan Darcourt were married, and Malina took a job with the Department of Agriculture, studying dust storms. Tsien was completing his doctorate and had no time for experiments, while Apollo Smith, unable to keep working for no money, left the Suicide Squad for an engineering job at Douglas Aircraft. Parsons and Forman picked up their old jobs at the Halifax Powder Company. In a final grim omen Weld Arnold, their mysterious patron, left Caltech and "completely vanished" from sight, never to be seen by the rocketeers again.

No one was more despondent than Parsons. Rockets were his life and the Suicide Squad had been his family. Getting to the moon was his one wish. His enthusiasm was childlike in its directness. Even as the rest of the Suicide Squad drifted away, he continued to mix powders and plan new rockets after he came home from work. The others' lack of enthusiasm, no matter how reasonable, must have been galling for him. Writing home to his parents, Malina admitted guiltily that "rockets have been having another sleep . . . Parsons is doing some experimenting with powder and is disgusted with me for not putting in more time to the research."

Parsons held a Halloween party that fall, on the same night Orson Welles gave his nationwide broadcast of H. G. Wells' *The War of the Worlds*. Perhaps he and his friends listened as Welles told in stentorian tones how "across an immense ethereal gulf, minds that are to our minds as ours are to the beasts in the jungle, intellects vast, cool and unsympathetic, regarded this earth with envious eyes and slowly and surely drew their plans against us." For the next hour, in a series of realistic-sounding news announcements, the nation was informed of gas eruptions taking place on Mars and astronomers' confusion at what they might portend, of the arrival in Grovers Mill, New Jersey, of a thirty-yard-long cylinder that hissed through the sky like a "Fourth of July Rocket." The invasion of the earth by Martians had begun. The public reaction, especially on the East Coast where the alien landings supposedly occurred, was one of mass hysteria; hundreds telephoned the police and fled their homes. One unfortunate resident of the actual town of Grovers Mill committed suicide rather than suffer a fate worse than death at Martian hands. If Parsons was indeed listening, he must have worn a rueful smile. For if hundreds of thousands of people could believe men from Mars had arrived by rocket to attack the earth, why would no one have faith in the rocketeers' modest attempts at getting to the moon? Why at a time when space travel and rocketry, including Parsons' own experiments, were making news, was the Suicide Squad on the brink of dissolution?

It was a taxing time for Parsons. The Suicide Squad had offered both comradeship and intellectual stimulus. Where could he find such solace again?

6

·············

THE MASS

*It is therefore a truism, almost a tautology, to say that
all magic is necessarily false and barren; for were it
ever to become true and fruitful, it would no longer
be magic but science.*

—SIR JAMES GEORGE FRAZER, *The Golden Bough*

*Any sufficiently advanced technology is indistinguishable
from magic*

—SIR ARTHUR C. CLARKE

A gong sounded. "We're all going upstairs," someone said. This
was how it usually began. Conversations drew to a close, and
Parsons followed the group that had been milling around the
living room out of the door. There were about fifteen men and
women, some as young as Parsons, a few others much older. An
air of excitement hung over the small group, like that of a theater
audience anxiously heading towards their seats before curtain.
The group's members seemed to be a mixture of actors, German
and Russian immigrants, and young bohemians. They made their
way up a large swooping staircase to the third floor, where a
single precipitous set of steps led through a trapdoor into the attic.
The attic room was small, rectangular, and dark, illuminated by a
single dim globe. The air was close with pungent incense that
came fuming from a brazen pot on the floor.

Wooden pews faced a black-and-white-checked stage. On
an altar swathed in black cloth stood a tablet carved with hiero-

glyphs, lit by a phalanx of twenty-two candles. A chalice sat beneath it, surrounded by roses. On either side of the altar were two tall obelisks—one black, one white. A black sarcophagus dominated the stage. A gauze curtain hung over its front, hiding its contents from view. A funereal dirge sounded from an organ at the side of the stage. The room, now full, went quiet. A sallow man in a white robe and sandals walked forward carrying a book, holding it in front of him reverentially. Bowing before the altar, he kissed the book three times, opened it, and placed it in front of him. He turned and in a voice of surprising stridency proclaimed: "Do what thou wilt shall be the whole of the Law." From all around Parsons the crowd responded, "Love is the law, love under will."

An imposingly large woman appeared onstage, clutching a sword in the manner of a Valkyrie. "The Priestess," someone whispered to Parsons. "The Virgin," someone else sniggered. Striding towards the sarcophagus, she cut through the covering veil with her sword. Out of it walked a short, bony man, carrying a lance close up against his chest. He wore a red velvet cape and on his head a diadem in the shape of a snake. His voice thin and piercing and marked by a distinct English accent, he began to intone, "I, Priest and King, take thee, Virgin pure without spot."

The attic in which Parsons found himself was nestled under the roof of a large wooden house on Winona Boulevard in a run-down neighborhood of Hollywood. Parsons had been persuaded to drive down from Pasadena by two social acquaintances of his, John Baxter, a gay man who was particularly fond of Jack, and his sister Frances, a lesbian, who had taken a shine to Helen. They were irregular attendees of what they called the Gnostic Mass of the Church of Thelema. It is possible that Parsons' curiosity about the mass was aroused when he learned that it was the creation of an English writer and magician, little known in America, named Aleister Crowley; for by a curious coincidence Parsons was already familiar with Crowley's work. Months prior to his visit to Winona Boulevard, he had called on his friend Robert Rypinski (the owner of the Pasadena used-car lot), and while browsing through his bookshelf, Parsons discovered a book named *Konx Om Pax,* dating from 1907. Quotes from Catullus, Sappho, the Koran, and Dante peppered its opening pages, while the rest of the book was filled with cabbalistic fairy tales, Chinese characters, and strange meandering rants. Written by Crowley, it was a puzzle book, which must have engaged Parsons' literary dilettantism, but it also spoke of a "hidden knowledge," of mystics and demons, of magic. For Rypinski the book had been little more than a curiosity, but for Parsons it seemed to be a revelation. "I don't know what it [the book] meant," remembered Rypinski, "but this was like real water to a thirsty man to Jack."

Since his childhood attempts to conjure up the devil, Parsons had always been attracted to tales of the hidden, magical mysteries that lurked behind the "real" world. He had always been able to picture Avalon and El Dorado amidst the slag heaps and corrugated-iron sheds of the powder companies and imagine "alchemistic doors" opening overhead when most others would have seen a hazy sunset. The copy of *Konx Om Pax* that spoke to Parsons so directly had seemed as perfect a starting point for the study of these hidden magical worlds as Robert Goddard's much lambasted *A Method of Reaching Extreme Altitudes* had been for his rocket work.

Back in the heavy, airless atmosphere of the attic on Winona Boulevard, the ancient plot of ritual continued. The ceremony combined quasi-Masonic gestures with hints of the Catholic mass, but an intense sexual motif made it unique. The priestess stroked the priest's lance and the priest kissed the priestess between the breasts. They made lavishly sensual declarations in the manner of high Victorian aesthetes: "I love you! I yearn to you! Pale or Purple, veiled or voluptuous, I who am all pleasure and purple, and drunkenness of the innermost sense, desire

you . . . Burn to me perfumes! Wear to me jewels! Drink to me . . .
I am the blue-lidded daughter of sunset; I am the naked brilliance
of the voluptuous night-sky. To me! To me!" The priest handed
out goblets of wine and "cakes of light," thicker and more ro-
bust than normal communion wafers, and read out a long and
obscure list of "saints." Fictional characters tripped fast on the
heels of authors; gods were paired with men. Dionysus followed
Lao-tzu, the founder of Taoism. Pan preceded Pythagoras. The
Victorian poet Swinburne came fast after Heracles. However,
the most approbation was saved for the last name mentioned—
that of Sir Aleister Crowley. Finally, the priest announced to the
congregation, "The Lord bring you to the accomplishment of
your true Wills, the Great Work, the Summum Bonum, True
Wisdom and Perfect Happiness." Within thirty minutes of starting the mass was over.

The congregation made its way downstairs. "We make the
cakes of light with animal blood," Parsons was told; "they
should really be made with menstrual blood." The air was still
charged with sexual tension, and couples began to pair off to
different corners of the cavernous house. The Baxters presented
Parsons to the three senior members. The woman who acted as
the indomitable virgin in the mass introduced herself as Regina
Kahl. She was forty-eight years old, an amateur opera singer
and a drama teacher at the nearby Los Angeles City College. A
few years before, she had gained mild infamy while acting in a
production of Aristophanes' *Lysistrata* that had been raided by
the Los Angeles Vice Squad. She had spent the night in jail for
performing in the ancient Greek play.

Next was Jane Wolfe, sixty-four years old and looking worn
and harried. In her youth she had been a silent film actress of
some renown, best known for playing Mary Pickford's mother
in the 1917 film, *Rebecca of Sunnybrook Farm*. She had lived in
Sicily with Crowley and smoked opium with him, and she forever looked back on this time with deep nostalgia. Then the
gaunt man who had acted as the priest in the ritual approached

the group. Away from the attic his stature was diminished somewhat. His headpiece had concealed the fact that he was bald,
and now Parsons could see that the robe he wore had been made
from a piece of old theater curtain. But behind his large bony
nose and strong jaw flickered an indefinable charisma. He introduced himself as Wilfred T. Smith, fifty-three years old and a
clerk for the Southern California Gas Company. He was born in
Tonbridge, England, where his father had been the sixteen-year-
old son of a prominent English family, his mother a maid in the
household. The details of his parentage were hushed up, and
Smith had lived with his grandmother before being persuaded
to travel to Canada once he turned twenty-one, a common fate
for social misfits at the time. There he had become familiar with
Crowley's teachings, and he had even met the man in person.
He could quote Crowley's poetry from memory, and he talked
to Parsons with intelligence and interest, offering to lend him
some of Crowley's other books. He told Parsons that the mass
was held every week if he wanted to attend. Or, he suggested,
Parsons might first want to come to one of the secular social
events held at the house—the annual Walt Whitman birthday
party, for instance.

If it was a community Parsons had been searching for, he
had found one. But he was initially hesitant about joining it.
From this first meeting in January 1939, Parsons would spend
over a year attending sporadically, mulling over the mixed emotions he felt about Smith, towards whom he would later admit
he felt both "repulsion and attraction," and about the group itself. But whatever Parsons' misgivings at returning to this strange
new society, it seemed his most likely hope if he was to understand and explore the work and magic of Aleister Crowley.

"Explain me the riddle of this man," wrote one of Aleister Crowley's earliest biographers, and many have similarly struggled to
capture the multifarious bizarreness of the man who referred to
himself as Baphomet, Frater Perdurabo, The Master Therion,

666, and The Great Beast. A novelist, poet, philosopher, mountaineer, chess master, painter, big game hunter, but above all, magician, Crowley was born in England in 1875, into a wealthy but austere religious family. As a child he was allowed to read only the Bible; it did not convince him. He grew up to rail furiously against both British prudery and the strictures of orthodox religion, proudly positioning himself in direct opposition to God and the established Christian church. To this end he identified himself with the "Great Beast" from the book of Revelation.

Backed by a sizeable inheritance, Crowley set himself the task of becoming an adventurer on both the physical and mental planes. In 1902 he took part in the first attempt to scale K2 in northern Kashmir, then the highest mountain accessible to Europeans in the world. Although he failed to reach the peak, possibly because of his insistence on carrying his traveling library with him, he and his companions climbed to a record altitude of 21,000 feet. His mountaineering career, however, would be ended just a few years later when he led a disastrous attempt on the holy mountain of Kanchenjunga on the Nepal/Sikkim border, the third highest peak in the world. Crowley narrowly escaped being swept to his death in an avalanche, and his name was somewhat blackened when he was accused of ignoring his fellow climbers' cries for help.

As well as mountaineering Crowley devoted himself to painting and literature, writing reams of poems and verse dramas in a high Victorian style—if not with a high Victorian content. G. K. Chesterton praised his early writings, but many others detested or just ignored them. Laced with blasphemy and, at times, shockingly pornographic, his self-published books were regularly impounded by customs. His prose writings were as unconventional as his poetry. He wrote a systematic study of narcotics and their usage in his book *Diary of a Drug Fiend,* and he was one of the earliest western proponents of yoga, on which he wrote his tract *Yoga for Yahoos,* later retitled *Eight Lectures on Yoga.*

Crowley was an obsessive self-promoter—even faking his own death to drum up interest in one of his painting exhibitions—and he turned his hand to anything that interested him or shocked others, the two usually being one and the same. He played at theatrical impresario, leading the "Ragged Ragtime Girls," an all-girl string septet, on a disastrous tour of Russia. On a trip to the United States he declared himself an Irish Nationalist and called for the dissolution of the United Kingdom, and as the First World War raged he turned his hand to writing pro-German propaganda. By the early 1930s he was quasi-employed as a spy for the British Secret Service in Weimar Berlin, although this merely seemed to offer him the opportunity to troll through the city looking for male prostitutes in the company of the writers Stephen Spender and Christopher Isherwood.

He became an infamous character in both London and Parisian societies with his shaven skull and piercing eyes. Possessed of a scatological wit and "baulking erudition," he courted slander and picked fights in public. He was pompous and a snob, laying claim to a nonexistent knighthood and styling himself at various different times as Lord of Boleskine, a Highland laird, Count Vladimir Svareff, a young Russian nobleman, and Prince Chioa Khan, a Persian prince, not to mention the many previous incarnations he also claimed. He delighted in discussing his sexual peccadilloes, an outrageous thing to do at the time, and his overpowering, mesmeric personality often damaged those around him, particularly the women who flitted in and out of his life. Alcoholism and the asylum were not uncommon ends for those who had known him too well. The British tabloids labeled him "The Wickedest Man in the World," a role that he inhabited with pride. But even the writer Somerset Maugham, who had known Crowley in Paris and taken an immediate dislike to him, could not condemn him absolutely. Maugham described him as "a fake, but not entirely a fake," and made him the basis of his early novel *The Magician*. Crowley's glimmer of authenticity stemmed from the fact that for all his polymathic and, some might say, psychopathic tendencies, he held firmly to two underlying and quite genuine ambitions: first, his wish to practice what he called *magick,* "the Science and Art of causing Change to occur in conformity with Will," and second, the establishment of his own self-coined religion of "Thelema."

Mystical societies were hugely popular in late nineteenth century Britain. The net of empire had dragged back not only material plunder from the East but also a vast number of alternative religious practices—Buddhism, Hinduism, and Zoro astrianism. Such works of comparative religion as Sir James George Frazer's *The Golden Bough* had fostered an increasing interest in esoteric thought within middle-class Victorian circles and prompted a resurgence of such secret fraternities as Rosicrucianism and Freemasonry. Crowley joined one of the most influential of these societies—The Hermetic Order of the Golden Dawn—whose members were drawn from all levels of British society. The poet W. B. Yeats was a notable adherent. It was the Golden Dawn's belief that humans were only partway up the ladder of psychical evolution and that, if properly disciplined, the human will was capable of anything it wished—most notably contacting intelligences that existed outside the physical world. To communicate with these beings, members performed rituals intended to alter their consciousness and to allow them to attune their minds, rather like a radio set, to the different worlds surrounding them. The exact details of these rituals were the order's greatest secret. Only by working one's way up the strict hierarchy that governed the order—through a process of study and financial donation—could one attain them. For Crowley, this ascension was as thrilling a challenge as climbing any mountain. With his natural intensity and eager scholarship, he progressed through the ranks as rapidly as the laws governing the order would allow him. Indeed, impatient for results, he began conducting, and bragging about, magical rituals that were far in advance of his actual standing within the magical order.

His precociousness and arrogance seemed to have infuriated the other members, who moved to block his rapid ascent. Frustrated, but with an undiminished appetite, Crowley soon found a way in which he could answer to no one but himself.

Crowley married his first wife, Rose Kelly, in 1903. He dubbed her his Scarlet Woman, a term he used for women who acted as his personal spiritual mediums. Rose traveled with him to Cairo in 1904, where Crowley was preparing his latest adventure, an undercover journey to Mecca (which he was forbidden, as a non-Muslim, to enter) in the footsteps of his hero, the scholar, explorer, and Orientalist Sir Richard Burton. Once there, Rose fell into a hashish-induced trance and announced to Crowley that Horus—the falcon-headed god of ancient Egyptian mythology—was waiting for him. Rose told him that he was to go to the temple he had constructed in their Cairo apartment for his magick rituals. Crowley obeyed his wife's commands and, according to his own account, in the temple he heard a man's voice begin to speak from over his shoulder. Crowley wrote down every word the voice said, and by the time it had finished speaking, he had written what he entitled *The Book of the Law*. It was this book that played such a prominent role in the ceremony that Parsons watched in the attic of Winona Boulevard.

The book expounded the religion, or law, of Thelema, after the Greek word for "will." It took the form of a prose poem in three chapters, announcing the coming of a new age, the Aeon of Horus—the age of the child—which would commence, coincidentally, that very year. Crowley declared himself its prophet—the Logos (word) of the Aeon—and he asserted that his religion was destined to overthrow all established faiths. The central tenet of his book was the doctrine of total self-fulfillment; he expressed this creed with the phrase, "Do what thou wilt shall be the whole of the Law."

Ironically, it would be this central creed that would forever hamper greater acceptance of his religion, for the doctrine of

Thelema appeared to the press and public alike as little more than a call to debauchery. Throughout his life, Crowley would insist that he never intended it to be read as such. Rather, it was a call to self-awareness, to the discovery and pursuance of the follower's true will. But the phrase's ambiguity seemed somehow representative of Crowley's own personality. In both his lifestyle and writings, he was a mesmerizingly inconstant man. Just as he chafed at society's restrictions, so he subverted his own decrees when it suited his purpose, often to the profound consternation of his disciples, who treated him with something approaching papal infallibility. Indeed, his quicksilver character also helped ward off those followers who thought they could supplant him. Although he would write, "My sword to him that can get it," Crowley's changeability meant that any attempt on his leadership was doomed to failure. He made the rules and the rules allowed no one to displace him. He was perhaps most inconsistent when it came to sex, abhorring the term "free love" even as he practiced it to its fullest extent. But while many of his biographers have tried to simplify his character—calling him either a shameless debauchee or a misunderstood holy man—most have judged Crowley's belief in the *Book of the Law* as a profound religious text to be sincere, if infuriatingly irregular.

After his fateful vision, the dissemination of Thelema became the "Great Work" of Crowley's life. Upon leaving the Golden Dawn, he began his own secret society called the A∴ A∴ (also known as the Argenteum Astrum or Silver Star) in which he offered individuals personal instruction in the mysteries of magick. He came to realize, however, that if his Law of Thelema was to be spread to a wider audience, he could not rely on a secret society. He needed an open outlet for his teachings, not to mention an inlet for donations. In 1912 he had joined a small quasi-Masonic organization named the Ordo Templi Orientis, or OTO, which boasted 500 members spread across Germany, Austria, and Switzerland. Crowley seized control of the OTO, started a chapter in Britain, and began rewriting its rituals, grafting *The*

Book of the Law into the society's texts and adding the formally choreographed Gnostic Mass as its ceremonial centerpiece. One OTO ritual he preserved, however, was the use of sex as an intrinsic component of the working of magick. The OTO's members believed that sexual ecstasy could lift one to a different plane of consciousness, a common concept in Eastern religion. But the sexual act had to be controlled, so that it became a further extension of meditation. Followers would pay Crowley an annual subscription, as well as fees each time they progressed along the ten degrees of attainment. (Perhaps as a means of encouragement, the secrets to sex magick would only be revealed once the ninth degree was attained.) Thus it was through the OTO, and its amalgam of magickal practices, that Crowley began to propound his religion of Thelema across the world.

By 1939, the year of Parsons' first attendance at the mass on Winona Boulevard, Crowley was sixty-four years old and no longer the devil of his youth. A life spent exploring the unconscious with the aid of every possible stimulant, sexual practice, and magick ritual had tired his body and his finances. The novelist Anthony Powell remembered meeting Crowley in his latter years. His face was "dull yellow in complexion, the features strangely caught together within the midst of a large elliptical area, like those of a horrible baby, the skin of porous texture, much mottled, perhaps from persistent use of drugs."

Living in a small flat in London's Jermyn Street, Crowley was cared for by an ever decreasing circle of friends and acquaintances. His diaries from the time record the pains of a body breaking apart at the seams. Dental catastrophes, asthmatic attacks, fainting spells, and insomnia were compounded by fevers, sweating, constipation, and diarrhea. He was by now a heroin addict; the habit helped ease his pains but threw him close to death whenever his daily dose of "tri" did not appear. Despite his infirmities he kept up his libidinous ways, and his charisma, if somewhat diminished, was still affecting to his lady

visitors. His diaries speak of hours spent performing cunnilin-
gus, of sleeping with female prostitutes and picking up bus con-
ductresses, interspersed with roars against hateful impotence.
"Alice here," reads one entry; "I had severe Freudian reluctance
to do any thing (v. sick after tea) But frigged her for human kindness' sake." Besides heroin and sex, chess and the cinema appeared to be the greatest pleasures left to him. He writes of
watching *A Night to Remember* (a film about the sinking of the
Titanic) four times, and describes his reaction to the cartoons
screened before the feature films: "Donald Duck's Garden tech-
nicolour: infinitely sad!"

His health was not improved by his huge workload. Crowley was still turning out a prodigious amount of poems, plays,
Thelemic writings, and trying to raise the money to publish
them. His mind was still clear, and his tireless devotion to his
Great Work was undiminished. But more and more it seemed as
if the world had tired of him. The OTO, five hundred strong at
its height, had shrunk to less than fifty members scattered across
the globe, and the occasional donations he received were immediately spent on maintaining his grandiose lifestyle, bankrupting
him still further. His last significant sally into the spotlight took
place in 1934, when he brought libel proceedings against a publisher who had claimed he was a "black magician." The judge,
however, decreed that it was virtually impossible to libel a man
of Crowley's reputation. The case bankrupted him.

Despite his best efforts, his notoriety had never translated
into attracting members to the OTO. The heyday of the mystical society seemed to have faded. Crowley was an anachronism,
a bygone tabloid devil. "No cash, no credit, no news, no tobacco, no friends, no printer, no hope, no bloody nothing," he
lamented to his diary. But there was one hope left—America.

If ever there had been a place to begin a religion, it was Los Angeles in the first half of the twentieth century. The attic chants
that Parsons had joined on Winona Boulevard were part of a

hubbub, a bedlam of prayer that emanated from the city. If a deity or deities had been listening, they would have been deafened by the white noise. The orange groves had been uprooted and replaced by cheap bungalows, and it was the boosters for religion who now prospered in the sun. A large, displaced populace was particularly responsive to a host of cults and cranks, enticed by both religious promises and the carrot of community. As Crowley wrote to Wilfred Smith at the height of the depression, "The world is drowning—that is exactly why it will clutch at a straw."

The nineteenth century had seen all manner of new religions rise and fall in the United States. A fascination with what the religious historian Sydney Ahlstrom called "harmonialism"—a belief that spiritual, physical, and even economic well-being flow from a person's connection with metaphysical forces of the cosmos—manifested itself in such new forms of thought as Spiritualism, Christian Science, New Thought, and Theosophy. These radical religions abandoned much traditional Christian doctrine and prospered on a mixture of charismatic founders, complex internal hierarchies, secret doctrines, and elaborate rituals. Like the mystical societies of Victorian Britain, many of these new religions incorporated aspects of Eastern religions. These had been brought to the United States' attention not only by the Japanese and Chinese immigrants who had flocked to the Pacific Coast, but by Hindu missionaries who had appeared at the World Parliament of Religions in Chicago in 1893. While they introduced the American people to such new words as *reincarnation, nirvana,* and *Karma,* the new religions also echoed the creed of self-reliance that had been an article of faith in American religion and culture for almost a century. This "new age in religion" found particularly strong footing in Los Angeles, where an aspirational population desperately sought a philosophy that offered the spirit what California had offered the body. Drawn by the promises of instant gratification inherent in the gold and health rushes that had prompted their exodus west, Angelenos expected the same from their religions.

Such a powerful demand was met with alacrity. The Church of Light disseminated its "Religion of the Stars" through classes preaching the use of tarot cards and astrology. The Institute of Mentalphysics, founded by Edwin Dingle, a former student of Tibetan monks and author of *Breathing Your Way to Youth,* offered to teach secret Oriental laws through its correspondence course. Just off Hollywood Boulevard, the lotus-scented Vedanta Society told how to transcend the limitations of self-identity through the study of ancient Hindu scriptures. Authors Aldous Huxley and Christopher Isherwood would later seek instruction in its techniques. Few establishments were as grandiose as that of the Ancient Mystical Order Rosae Crucis (AMORC), whose Egyptian museum in San Jose took up an entire city block. It stressed the virtues of reason and science while also suggesting that ancient Egyptian wisdom would allow its followers to release the hidden powers inherent in man.

A number of religious groups assimilated the new wonders of science into their teachings. The Superet Light Church, founded in 1925 by the "atoms aura scientist" Josephine C. Trust, taught of good and evil atoms, invisible spectrums and the spiritual significance of favorite colors. Others focused on direct individual improvement in health. The Church of Divine Science preached the gospel of "perfect action and perfect thinking . . . perfect breathing and perfect circulation, perfect digestion and perfect generation, perfect voice and perfect speaking."

There was a multitude of movements to choose from—the "I Am," Mankind United, the New Thought Alliance, the Christ-Way College of Occult Science, and The Occult Science of Christ Church. Even fundamentalist forms of Christianity, such as that preached by Aimee Semple McPherson's International Church of the Foursquare Gospel at the giant Angelus Temple in Echo Park, seemed to have been warped by the Los

Angeles sun. McPherson created a flamboyant show business atmosphere around her evangelical church. She preached sermons over the radio—she was the first woman to be awarded a radio broadcast license in the United States—and enthused the 5,000 strong congregations she regularly attracted with contemporary music, bizarre pageantry—ministers dressed in armor, brandishing swords—and extraordinary entrances, once driving down the aisle of her temple on a motorbike. Her church's message spread across the nation, gaining her some 80,000 members, but its leader proved increasingly unstable and prone to scandal as she suffered a string of nervous breakdowns and unseemly divorces, and eventually faked her own kidnapping. She would die from an overdose of barbiturates in 1944.

Such a high concentration of new and unorthodox religions was not to everyone's taste. Los Angeles "swarmed with swamis, spiritualists, Christian Scientists, crystal-gazers and the allied necromancers," spat H. L. Mencken when he visited the city.

Although Jack Parsons was brought up in an apparently traditional Protestant home, he was no stranger to the alternative religions bubbling around him. Along with thousands of other Angelenos, he had taken a passing interest in one of the most mainstream and highly influential of the new movements—Theosophy. The Theosophical Society was founded by the fraudulent psychic Madame Blavatsky in 1875. Its adherents claimed to be the guardians of ancient wisdom which they had obtained from a secretive brotherhood of "highly evolved adepts and masters" who lived in the Himalayan Mountains. Theosophy was both a philosophy and a religion, preaching the doctrine of reincarnation as well as spiritual evolution. Through the study of its teachings, one could ascend through the astral plane to an understanding of the divine. An obscure fourteen-year-old Indian boy named Jiddu Krishnamurti had been dubbed the movement's new Messiah or "World Teacher," but he had renounced these claims in 1929 and now lived in Ojai, California, eighty miles north of Los Angeles, where he taught and gave lectures. In the years immediately before he attended the Gnostic Mass, Parsons and Helen traveled up the coast a number of times to hear Krishnamurti talk. But while it seems that Parsons was interested, he was never completely convinced. He would later write of being "nauseated" by Theosophy's talk of the "good and the true."

Perhaps no other book shaped Parsons' interest in myth, magic, and alternative beliefs more than Sir James George Frazer's *The Golden Bough*. Parsons would refer to it in his letters throughout his life, and he often recommended it to those newly interested in the occult—that is, the supernatural world and its manifestations in magic, alchemy, astrology, and other secret or mysterious arts.

Frazer was an anthropologist and fellow of Trinity College, Cambridge. He wrote *The Golden Bough,* first published in 1890, after studying the mythologies and belief systems of the world's cultures. In the thirteen volumes which make up the work (a condensed one-volume edition was also issued), Frazer juxtaposes African tribal rites with the practices of Egyptian cults, Greek mythology with rural British folklore. His startling and controversial hypothesis was that the history of human culture demonstrated an evolutionary development of thought—from magic to religion and finally to science. Frazer became one of the first scholars to treat magic not as blasphemy or heresy but as a legitimate (if flawed) system of thought. Frazer's book highlighted the parallels between the rites, superstitions, and magic of primitive and pagan cultures and those of Christianity and orthodox religion. He suggested, for example, that the biblical story of Christ's resurrection recreated the pagan celebrations of spring. His work made a vast range of primitive customs suddenly intelligible to a Western audience. *The Golden Bough* was popular with both scholars and laymen, and it dramatically influenced the work of Sigmund Freud and Carl Jung—Frazer's depiction of tales of myth and romance as echoes of ancient rituals chimed with Jung's description of archetypes that exist within the collective unconscious—as well as of writers such as T. S. Eliot, whose poem *The Waste Land* resounds with its ideas. Frazer's suggestion that powerful manifest forces, more primal than Christianity, linked mankind through time and space must have been highly seductive for a man like Parsons, who was already a keen searcher for the mystical in everyday life.

While Frazer doubted magic's efficacy, calling it "a mistaken association of ideas," he saw that in its fundamental conception, ritual magic was strikingly similar to science. Magic, like science, was an attempt to control events by performing technical acts. Although Frazer believed that the reasoning behind magic was faulty, its basic methodology was not. He wrote with interest of Scottish witches who beat a "wind-stone" to create storms and of Papuans who believed that a man's sweat could be used by his enemies to cast an enchantment over him. To Frazer these acts of magic were the logical precursors to scientific experiment, for as with science "the succession of events is assumed to be perfectly regular and certain, being determined by immutable laws, the operation of which can be foreseen and calculated precisely." Both magic and science, he wrote, "open up a seemingly boundless vista of possibilities to him who knows the causes of things and can touch the secret springs that set in motion the vast and intricate mechanism of the world." Magic, with its unshakeable belief in cause and effect, made it "the bastard sister of science." Parsons would always treat magic as such, seeing it as a strictly literal branch of learning, one that could be mastered by concentrated scientific application.

Despite his unconventional character, Aleister Crowley had not taken to the wild variety of Los Angeles. He had visited it only once while on a tour of the United States and Canada in 1915–16, when he had described it, without a hint of insincerity, as populated by "the cinema crowd of cocaine-crazed, sexual lunatics, and the swarming maggots of near-occultists." It was while visiting the Vancouver branch of the OTO on this same tour that he had met Wilfred T. Smith, the man Parsons recently encountered at the mass on Winona Boulevard. A correspondence had resulted, with Smith sending Crowley money and Crowley advising him on magick and advancing Smith through the ranks of the OTO. When Smith announced that he was moving to Los Angeles, he asked Crowley's permission to form a branch of the OTO, under the name Agape Lodge (agape being the Greek for "brotherly love"). Crowley acceded with glee. He was all too aware of the success of Aimee Semple McPherson and the huge audience and fortune she had gained for herself. Despite his personal dislike for the city, he hoped that there Smith might be able to capture a most valuable commodity, an "RMW"—Rich Man from the West—to bankroll his future projects. "Start an entirely new *habit of life*," he advised Smith in 1928. "Ingratiate yourself with sane people, people of importance. Put over the Law of Thelema as a spiritual, social and political movement. Get men interested in its economic advantages first of all—and so on ... The money's right under your nose, and you need only to get in with the right people to have it handed to you on a tray."

The Church of Thelema at 1746 North Winona Boulevard was registered and incorporated by Smith in 1934. Its neighborhood was poor but already religious: the house was located near both the Vedanta Institute and a nunnery. Following the incorporation Smith made a renewed push for publicity, taking advantage of Crowley's infamy in Los Angeles: A book of his erotic poems, *White Stains,* had been found at the murder site of the film producer William Desmond Taylor in 1922. Smith wrote to Crowley breathlessly: "Twenty seven attended *(the Mass)* last Sunday ... The Crowley Night was most successful, and an enthusiastic crowd of 150 came. 137 signed the guest book. We are gradually building up a mailing list."

The actor John Carradine visited the house, although it is difficult to say whether he came because of sincere belief or merely for research purposes: He was soon to play the organist of a satanic cult in the film *The Black Cat*. His fame was the exception. Crowley wanted the Los Angeles branch of the OTO to be attracting "great bankers and captains of industry," but the somewhat modest and self-conscious Smith could never clinch a substantial benefactor to bankroll Crowley's work and lifestyle. Instead, he attracted "communists and pacifists, or both."

One of these was Harry Hay, a young actor and communist who had been performing in Clifford Odets' play on unionization, *Waiting for Lefty,* at the Hollywood Guild Theatre. Hay would later become father of the gay rights movement in America, but he was hired to play the organ for the OTO's Gnostic Mass, having been drawn to the temple through his friendship with Regina Kahl. Crowley was known to have created homosexual sex magick rituals, a risqué idea at a time when homosexuality was illegal. The OTO also offered a safe meeting place when Los Angeles' gay bars were being ruthlessly targeted by the police, who had plenty of time on their hands since the end of Prohibition in 1933. Thus, many gay men were attracted to the OTO, with even Wilfred Smith, who often performed "exorcisms" on attractive male aspirants, joining in at times.

The OTO was as appealing for the rare sexual freedom it offered as for its role as a religious organization. While Los Angeles may have had one of the America's highest divorce rates, the general mood, fostered by the powerful antivice movements of the evangelical churches, made any organized challenges to the traditional family unit a rare phenomenon. The OTO's rituals contained a stronger sexual component than anything else being practiced in the city, and at times the high jinks threatened to overpower the main purpose of the group—the study of Crowley's work. Certainly, the mass was not always given the respect Crowley would have desired. Harry Hay would himself undermine the seriousness of the atmosphere by mischievously slipping such ditties as "Barnacle Bill the Sailor" and "Yes, We Have No Bananas!" (slowed to dirgelike tempo) into the contrapuntal themes he was asked to play.

After the initial successes about which Smith had bragged to Crowley, attendance at the masses fell. While the social parties continued to draw many people to the house, largely thanks to Regina Kahl's contacts in the acting world, the practicing membership of the OTO was reduced to a hardcore following of around ten people, some of whom lived within the house on Winona itself and shared the maintenance and costs of communal living. They were an unusual group. Louis Culling was a veteran of the First World War and a former cinema organist with a longtime interest in the occult. Max Schneider was a Prussian-born jeweler and astrologer. Oliver Jacobi was an employee of the gas company that Smith worked for. Carrie Dinsmore was a convert from the New Thought movement and inventor of the "slip-knot garment hanger" (30 percent of the royalties of which she donated to Crowley). Dr. George Liebling, who stayed in the house for a brief period, had been a protégé of the composer Franz Liszt and was a famed concert pianist. Perhaps the most well known of the OTO's members in local circles was Roy Leffingwell, a pianist and composer who was also known as "Pasadena's Greatest Booster" from his role as the official announcer of the city's annual Tournament of Roses.

On February 26, 1939, little over a month after Parsons' first visit to the OTO, a former Zeigfeld Follies dancer attending drama classes at Los Angeles City College was attacked on campus and died from her wounds. Her tutor had been Regina Kahl, the priestess at Agape Lodge. It did not take long for police investigators to discover the link between the murdered student and the meetings of the OTO at Winona Boulevard. The chief of police himself turned up to grill Smith, Kahl, and Wolfe about the suspicious goings-on in their house.

Although there was no connection between the girl's murder and the OTO, an avalanche of newspapermen soon descended on the house, keen to get the scoop on what had been termed the "Purple Cult." Headlines such as HIGH PRIESTESS IS COLLEGE TEACHER and PURPLE CULT RITES BARED sprawled across the newspapers. Smith and Kahl did their best to make the most of the publicity, inviting reporters to watch the mass and allowing pictures to be taken.

The house's guest registrar shows Parsons visiting only once more that year, for a party celebrating Smith's birthday. His interest in Thelema and Crowley, however, was growing. Smith judiciously began to sell him copies of Crowley's books, which he would study at home. Seeing her husband's mounting interest in Crowley, Helen also began to read the books. Although she was initially shocked by Crowley's sexist attitudes, she, too, was gradually drawn to what she read. By the end of 1939 Parsons owned a respectable collection of Crowley's work. The age of Horus, the wink of the grandeur of the past, even if it was wearing a robe made out of old theater curtain, seemed to be infinitely more in tune with his romantic leanings than Communism or even science fiction.

His enthusiasm was evident when he talked or corresponded with the other members of the OTO. He told them he was fascinated by Crowley's talk of hidden dimensions and access to forbidden planes. As a scientist he found that Crowley's magick teachings seemed to correlate with the work of "the 'quantum' field folks." The illogical nature of the newly coined quantum physics, in which the simple act of observation seemed to affect the physical world, and in which changes performed on one physical system could have an immediate effect on another quite unlinked system (the theory of nonlocality), would have seemed to Parsons to endorse the improbable possibilities of magic and especially the transformative powers of the magician himself.

Wilfred Smith wanted to initiate Parsons into the OTO as soon as possible, although Parsons' rocket work and Helen's initial reluctance meant the initiation would have to be delayed for a while. Nevertheless, Parsons' scientific learning, his natural aristocratic manner and his wealthy countenance all gave Smith hope that Parsons could be the "Rich Man of the West" he had been looking for all these years.

7

. . . .

Brave New World

The profession of magician, is one of the most perilous and arduous specialisations of the imagination. On the one hand there is the hostility of God and the police to be guarded against; on the other it is as difficult as music, as deep as poetry, as ingenious as stage-craft, as nervous as the manufacture of high explosives, and as delicate as the trade in narcotics.

—William Bolitho, *Twelve Against the Gods*

After ten long years of the depression, a new world was unveiled in April 1939 far from the blight of the past decade. "Building the World of Tomorrow" was the theme for the New York World's Fair, and even its location, on land that had previously been an ash dump, seemed to speak of phoenix-like beginnings. Towering over the fair were the Trylon and the Perisphere, a seven hundred-foot-high spire and an orb as wide as a city block. In their shade such astounding inventions as television, FM radio, fluorescent lighting, fax machines, Lucite, nylon, and 3-D movies were exhibited for the first time in public. Mile-long lines stretched out of the Futurama exhibition hall, a testament to the public's fascination with Westinghouse's walking, talking robot, Elektro. The fair promised a future of comfort, convenience and, above all, change. The concept of "suburbs" was introduced, as well as interstate highways, sleek new motorcars, and jet airplane travel. Sandwiched between the poverty of the depression and the imminent horrors of the Second World War, the fair was one of the greatest visionary exhibitions of the century.

The fair also proved the perfect backdrop for another gathering—the first World Science Fiction Convention. Around two hundred fans from across the country made the trip to New York, among them Forrest Ackerman and Ray Bradbury of the LASFL, both of whom had traveled 3,000 miles by train to be there. The convention featured some of the most famous science fiction writers of the day, including Jack Williamson and L. Sprague de Camp, as well as Frank R. Paul the preeminent science fiction illustrator of the time, who had drawn the first cover of Hugo Gernsback's *Amazing Stories* back in 1926. When Paul gave the keynote speech, entitled "Science Fiction: The Spirit of Youth," he took some well-aimed swipes at the old guard of scientific inquiry, including Caltech's head:

> The other day Dr. Robert Millikan said we should stop dreaming about atomic power and solar power ... As much as we love the doctor as one of our foremost scientists of the day, because he cannot see its realization or gets tired of research is no reason to give up hope that some scientist of the future might not attack the problem and ride it. What seems utterly impossible today may be commonplace tomorrow."

Also present was the science writer and former member of the German rocket society the VfR, Willy Ley, who had fled from Nazi Germany in 1935 as the German army had taken control of the rocket projects. Although he came to spur the young enthusiasts to imagine the very real possibilities of space travel, his attendance could not help but qualify the fans' boundless optimism in the future. A rampantly Fascist Germany had recently annexed Czechoslovakia; Italy had invaded Abyssinia. Britain, France, and the Soviet Union were squabbling ineffectually about the best way to block further Nazi expansion. As the world was being wracked by political ideologies, so the science fiction community had become riven by its own byzantine political struggles, as if mimicking the tumultuous events on the world stage. Two radically opposed fan organizations, the Futurians and New Fandom, had declared that they would be attending the convention. The politicized Futurians, whose ranks included a young Isaac Asimov, held that science fiction should rise to "a vision [of] a greater world, a greater future for the whole of mankind, and [should] utilize ... idealistic convictions for aid in a generally cooperative and diverse movement for the betterment of the world along democratic, impersonal, and unselfish lines." Opposed to them was New Fandom, the group that had organized the convention, who insisted that science fiction be read purely as entertainment. To them the Futurians were "dangerously red"; indeed, many Futurians were also members of the American Communist Party. Scuffles ensued and some Futurians were barred from entering the convention. For the apolitical Ackerman, who had bedecked himself in a silver Buck Rogers space suit to celebrate the occasion, "It was a moment of immortal sadness." As usual, science fiction was proving remarkably prescient: Soon science, and the Suicide Squad's rocketry project itself, would become swept up into the realm of politics.

Just months earlier, at a meeting of the Scientific Research Society, Sigma Xi, Frank Malina had given a talk which might have fit in at either the World's Fair or the science fiction convention. Entitled "Facts and Fancies of Rockets," it suggested that rockets strapped under the wings of a plane could greatly improve its performance. Following the lecture, Theodore von Kármán asked Malina if he would go to Washington, D.C., to give expert information on rocket propulsion to the National Academy of Sciences (NAS) Committee on Army Air Corps Research, on which Kármán sat. The NAS brought together committees of experts in all areas of scientific and technological endeavor to provide critical assistance to the government. Malina's trip was to be strictly secret.

General Henry "Hap" Arnold, Chief of the United States Army Air Corps and a close friend of Kármán's, had asked the NAS to study two critical problems of heavy bomber aircraft—the de-icing of windows on bombers flying at high altitude and the need to find something to assist the takeoff of heavily laden aircraft in combat zones, since sufficiently long runways would likely be hard to find in war zones. To Kármán, the Suicide Squad's rocket work seemed to offer the solution to this latter question.

Malina arrived in Washington, but he did not talk about rockets: He and Kármán had decided that because of the poor reputation of the word *rocket* in serious scientific circles, he should abandon it in favor of the word *jet*. Thus it was that the NAS heard of "jet propulsors" strapped under the bodies of planes to boost the speed, range, and takeoff time of aircraft. "Jets" would transform airplanes, allowing them to break free from the constraints imposed by the propeller. Naturally, Malina said coyly, he could not be too detailed about the exact benefits, as research was still in its infancy. However, if the committee would offer some financial backing, Malina was sure that he and the Rocket Research Group could help create record-breaking changes in aviation.

Seduced by both Malina's talk of "jet" planes and Kármán's wholehearted backing of the plan, the NAS gave the Suicide Squad $1,000 to prepare a proposal for a full-time research program into Jet-Assisted Take-Off (JATO) by June 1939. They were to show that it was possible to create rocket engines facilitating the "super-performance" of aircraft—in other words, that rockets, when attached to a plane, could shorten the time and distance of takeoff, increase the rate of climb, and increase the level-flight speed. The report was to be an extension of the work

they had been doing sporadically up until now and was to feature detailed studies on "jet propulsors" fueled by gas, liquid,
and solid propellants. Kármán would act as the project's guide.

Many still thought the project was a waste of time and
money: Jerome Hunsaker of the Massachusetts Institute of Technology, who had taken up the other problem of de-icing bomber
windshields, contemptuously told Kármán, "You can have the
Buck Rogers' job." But Malina, Parsons, and Forman were ecstatic. It was their first funding since Weld Arnold's mysterious
donation a year and a half earlier. Now it was another Arnold
who had come to their rescue, backed by the seemingly limitless
budget of the armed forces. The work of the past three years
was to be rewarded. "We could even expect," remembered an
amazed Malina, "to be paid for doing our rocket research!"

Parsons and Forman would work full time for $200 a
month—double what they had earned at the powder companies—while Malina would put in half time and defer taking
his degree. The windfall would not only allow Parsons to "afford smoking ready-rolled cigarettes" again; it would also ease
the burden on Helen, who was working full-time to support
them both. Parsons, now twenty-five, and Forman and Malina,
both twenty-seven, became the United States first government-
sanctioned rocket group.

You can take the rocket scientist out of the Arroyo, but you can't
take the Arroyo out of the rocket scientist. Government sponsorship may have lent the rocketeers a veneer of respectability;
however, it was the same old Suicide Squad, now reduced to the
core trio of Parsons, Forman, and Malina, that continued research. The group's recklessness did not decrease, nor did the
new funds lessen the dangers of their work. Once again the Caltech campus resounded with the "unnerving explosions of Parsons' rockets," noted Kármán.

On one occasion in their hut outside the GALCIT building,
Parsons was testing a volatile ethylene-and-gaseous-oxygen

mixture as a possible liquid fuel. Malina was sitting on a stool, marking down the numbers on various gauges connected to the rocket, when he remembered that Kármán had asked him to deliver a typewriter to his home. It's unclear whether the removal of Malina's controlling presence provoked Parsons to push the apparatus too far, or whether the equipment simply had a structural fault, but shortly after Malina left, two huge explosions cracked through the monastic quiet of Caltech's buildings. Smoke billowed upwards and a crowd rushed out from the nearby buildings. They found blackened concrete, burnt sandbags, and a shaken Parsons and Forman clutching their heads. An oxygen line had caught fire and ignited the oxygen tank itself. When Malina returned from Kármán's, he saw that one of the gauges from the apparatus had been propelled deep into the concrete wall, exactly where his head would have been had he stayed. The incident was just the latest in a long line of lucky escapes. The apparatus was not so lucky. Damage worth $250 was done, one quarter of their grant, and some serious rebuilding was needed.

Parsons and Forman were working full time, arriving late in the morning and working late into the night. Soon they were able to show that a rocket providing "super-performance" for aircraft was well within their reach. When they delivered their paper to the NAS in June, they felt sufficiently confident to ask for a budget of $100,000 to fund further research and construction of the rockets themselves. But when Kármán took the report to Washington, he found that their confidence was not shared by the NAS; the old prejudice against rockets was still strong. The rocketeers' budget was set at $10,000, and at least one member of the board was flabbergasted at even this sum. "Do you honestly believe," Kármán was asked, "that the Air Corps should spend as much as $10,000 for such a thing as rockets?"

The rocketeers were quite happy to get this princely sum, even if it was less than they had asked for. For Parsons and especially

Malina, military sponsorship was no longer quite the Faustian contract it had once seemed. Not only had their ethical qualms been slowly eroded by four long years of temporary employment and poverty, but the rise of Fascism had given a different moral shade to their work. "We decided that we were going to use rocketry to defeat Fascists," recalled Malina. "We felt that socially responsible engineers and scientists at that time had a mission to perform." The fact that they had been asked to create rockets not as munitions but as aids for aircraft also helped. When Germany invaded Poland on September 1, 1939, instigating the Second World War, the rocketeers' task became all the more imperative.

Despite the fact that they were members of America's first university-based rocketry program, now referred to as "GALCIT Project Number One," Parsons and Forman had not endeared themselves to Caltech. An invisible barrier—thrown up by both their happy-go-lucky personalities and the infra dig nature of the rocket work itself—prevented them from ever being truly accepted into the body of Caltech's scientists. Jeanne Forman, Ed Forman's third wife, felt that the resentment stemmed from jealousy. "Here were these 'dumb' people who were getting honors [from the military]. How dare they, when they hadn't gone to university." Parsons and Forman's joy in explosions, their tendency to follow whims, and their contempt for safety procedures would forever set them apart from the scientific establishment, even as it eventually came to embrace rocketry. Even Malina never fully reconciled himself to the explosive nature of their experimentation. "Sometimes," Malina groaned, "they are like inventors, in the worst sense of the word."

An example of this prejudice reared its head at a formal dinner held for various staff members at the illustrious Athenaeum, the Caltech faculty club. With its giant fireplace, oriental rugs, soft lamplight, and air of extreme erudition, the Athenaeum, said one architectural critic, made one feel like a Nobel Prize winner just by entering it. When Parsons and Forman received invitations to the meal, they thought they were finally receiving their just deserts for years of unpaid research; when they discovered that Malina hadn't received one, they ribbed him endlessly about it. An invitation for Malina did eventually arrive the day before the dinner—but it was accompanied by a note saying that Parsons and Forman had only been sent invitations because of a clerical error. Even though Parsons was by now listed in the Caltech staff directory, the privilege of attending the dinner was apparently reserved for "proper" Caltech members. Even for the thick-skinned Parsons and Forman, such a revelation would have been a slap in the face. Malina "didn't have the heart to tell them that they were invited by mistake" and so Parsons and Forman attended the black-tie dinner in high spirits, oblivious to the slight, and were "staff members for the night."

The Suicide Squad were now asked to deliver what they had promised in their report to NAS. The army air corps wanted a solid-fuel rocket that could deliver a constant, powerful thrust for at least ten seconds in order to give enough sustained lift to heavy bombers trundling down short runways. With his usual optimism, Parsons had felt certain that he could create a rocket to match these specifications. But there was a significant problem: As far as the rocketeers knew, the longest anybody had ever gotten a black powder rocket to burn was little more than five seconds. And even then the fuel did not burn steadily, producing thrust that was both weak and irregular. Parsons' confidence belied the fact that black powder rockets were still little more than fireworks, barely changed since their invention a millennium ago.

"A propellant needed to be found that could provide the proper thrust in a controlled manner," recalled Kármán in his memoirs. "This meant that the burning of the propellant in the rocket chamber had to proceed evenly, so that the pressure of the exhaust gases would not drop during the critical period of takeoff." Parsons realized that he had to create a new type of rocket, one with a propellant that would act as a slow burning charge. In theory it was quite simple. The rocket fuel needed to burn like a cigarette, from one end only (known as "restricted burning"), rather than being allowed to ignite on all sides instantaneously or burning along only one side of the rocket wall. If restricted burning could be achieved, depending on the dimensions of the "cigarette," any duration of stable thrust could be obtained.

Parsons knew he would need to create a fuel in the form of a "dense, tough, hard cylinder, completely free from cracks, which is pressed against the propulsor chamber [the rocket motor] wall to form a gas tight union with the wall." He began mixing low explosive black powder with high explosive smokeless powder. He then coated both the fuel and the inside of the rocket motor with various substances, even glue, to try to form a solid or liquid seal between the charge and the rocket motor walls. But the mixtures frequently cracked and formed fissures as they dried, enlarging the burning surface and causing uneven burns and explosions. He consulted explosives experts from the Halifax Powder Company, but they declared combustion of solid fuels in a rocket chamber to be inherently unstable.

In desperation Parsons even tried an idea that Jack Williamson, the science fiction author, had put forward in his 1939 story of space travel, "The Crucible of Power." The spaceship in the story consists of four separate parts or steps, described in great detail: "Each step containing thousands of cellules, each of which was a complete rocket motor with its own load of 'alumilloid' fuel, to be fired once and then detached." But when Parsons tried to build these multicellular solid-fuel rockets—pressing separate cartridges of fuel one after the other into the rocket chamber—either the cartridges did not light, or they lit all at once. (In fact, Williamson's story predicted this latter problem, which ultimately causes the crash that dooms his space travelers.)

"The new apparatus being designed is changed from day to day and the machinist is getting dizzy from instructions and counter-instructions," wrote Malina to his parents, as he and the rocketeers worked at a frantic pace on "GALCIT Project Number One." Explosions at Caltech were now par for the course. "We upset the whole campus. People will get used to it, I hope." While eating breakfast at home one morning, Malina heard a huge explosion emanating from the Institute. "At first I thought it was our apparatus again. I looked at my watch and saw it was 8:30, which I was sure was too early for Parsons and Forman to be at work, and I was correct." The explosion had actually been caused by improperly stored chemicals in the chemistry building

It was time for Theodore von Kármán to lend his immense theoretical talents to the problem. In the spring of 1940, having listened for months to the unremitting sound of Parsons' test rockets exploding, Kármán spent an evening writing down four differential equations describing the conditions necessary for a slow burning fuel to work in a rocket. The next day Kármán gave them to Malina, saying, "Let us work out the implications of these equations; if they show that the process of a restricted burning powder rocket is unstable, we will give up; but if they show that the process is stable, then we will tell Parsons to keep on trying."

Parsons' experimentation would now take a back seat to theory. Retreating into his study, Malina began analyzing Kármán's equations. Solving them would tell Malina whether their work was theoretically possible. In short, these four equations held the future of their research. Malina never mentioned how long it took to work out these equations, but they were by no means simple. As Dr. Benjamin Zibit, the historian of science has described, "The solution of the equations yielded an important insight: the direct relation between surface area of the burning propellant and the diameter of the nozzle throat. If the proportion of burning propellant did not exceed the capacity of the nozzle throat to channel the hot gas in a uniform flow, then burning would remain a stable process." However, if the volume of burning propellant did exceed the capacity of the nozzle throat, no matter how hard Parsons experimented, he would never be able to create a stable fuel. Malina eventually emerged triumphant. Kármán's equations had proved that the process was inherently stable. Parsons just needed to find the right combination of ingredients to create a fuel of suitable strength capable of uniform burning.

Parsons leapt into his work with renewed vigor and Kármán settled back to listen to more explosions. It was long, hard work, but Parsons remained joyously optimistic. Overflowing with enthusiasm, he scrawled a poem on the back of a piece of Caltech stationery:

> The path is hard and the night is long
> And the way is bleak and weary
> But my heart is high as [we] trudge along
> (Tho the road is long and dreary).

He tried every possible variation he could think of—different powder mixtures, various loading techniques—while Forman came up with a vast number of differing motor and nozzle designs in which to test them. They had to find exactly the right combination of powder, loading technique, and rocket motor design.

At this point, however, the Caltech authorities declared that they had finally had enough. If this work was going to continue endangering lives and concentration, it was not going to be done on campus grounds. There was only one thing for it: back to the Arroyo Seco, where the Suicide Squad had begun their rocket experiments four years ago. Leasing six acres of the western bank of the Arroyo Seco from the City of Pasadena, the rocketeers set about building two or three ramshackle wood and corrugated–sheet metal buildings to work in. Over the next months Parsons traveled to the Arroyo hundreds of times, pains-

takingly testing scores of different powder formulations in the
rickety and ill-lit buildings. It was not a comfortable process.
Wedged at the foot of the mountains, the Arroyo test site received few cooling breezes. Temperatures that summer were in
excess of a hundred degrees Fahrenheit, and the corrugated-iron
sheds amplified the already stifling heat. Since the rocketeers
were only separated from their test stands by a wall of used railroad ties—the cheapest of building materials—when the rocket
motors fired, the temperature soared even higher.

The rocketeers were now testing three separate rocket motor
apparatuses, and like battle-weary soldiers at war, they no
longer thought of an explosion as dangerous but merely as in-
convenient. In fact, they were becoming quite blasé about the
proceedings. "Today we made a test and had a nice 'blow.' No
serious damage, just about a week's worth of repair," wrote Malina to his parents.

Parsons, bare-chested and sweating profusely, crammed
himself into the tiny test buildings and sat taking notes from the
instruments. How long did this powder burn? What was its
maximum thrust? Most of the time it was hard to tell. The equipment being used was so basic that the only way to record data
was taking photographs of the gauges in the midst of the experiment. The cameras were calibrated to take between one and
four pictures a second. Unfortunately, since most tests lasted less
than a tenth of a second, the photographs showed either nothing or the blur of a needle across the whole dial.

Finally, Parsons made a breakthrough. With a black-powder
propellant of his own composition (roughly 72 percent potassium nitrate, 15.5 percent charcoal, and 12.5 per cent sulphur),
he suggested loading the rocket motor using a mechanical press.
The powder would be pressed down in small, one-inch incre-
ments at high pressure. It was a laborious job, but the process
would limit the amount of space between the grains of black

powder and thus insure steady burning. The steel rocket motor itself would be lined with blotting paper which would allow an even burn to proceed along the side of the motor. When they tested the setup, the burning proceeded down the length of the tube in a controlled and stable manner. It was the first time that such a restricted burning rocket had ever been made and "the group was jubilant."

By June 15, 1940, GALCIT Project Number One was able to report a number of positive results to the army air corps. In addition to his success with solid-fuel rockets, Parsons had also managed to find an alternative oxidizer for his liquid-fuel rockets. The air corps had declared liquid oxygen too impractical for combat situations since it had to be kept at extremely low temperatures. After months of testing chemicals in open crucibles, Parsons had stumbled upon red fuming nitric acid, a solution of nitric acid and nitrogen dioxide better known as RFNA. It was by no means an ideal answer—it was highly poisonous and very corrosive—and Parsons' trousers and hands had been burnt yellow and brown from the toxic vapors it released. Nevertheless, when combined in a rocket motor with fuel—a mixture Parsons had devised of gasoline, benzene, and linseed oil, among other materials—it provided a powerful thrust. The air corps doubled their solid-fuel rocket funding to $22,000 with the understanding that Parsons, Malina, and Forman would prepare flight tests, with the rockets actually strapped to an airplane, the following summer.

The extra money caused a spurt of activity which created new interest in their work, and the squad brought a few new faces on board. Martin Summerfield, a Caltech friend of Malina, added some mathematical clout to the group. A short, bespectacled boy from Brooklyn, New York, Summerfield had previously worked in the Caltech physics department, principally in optics, X rays, and infrared radiation. He had been Malina's roommate as well as a member of Sidney Weinbaum's Communist salons, and Malina saw that his open intelligence, not to mention his legendary absentmindedness, made him a suitable match. The rocket group also applied to the Work Progress Administration (WPA)—part of Franklin D. Roosevelt's New Deal program—for help in constructing new buildings, and they were sent eighteen workers. Meanwhile, other graduate students from Caltech began to gravitate towards the group's experimental work in the Arroyo. Even Clark Millikan, the head of the GALCIT wind tunnel who had derided their work early on, was beginning to show an interest in the rocket's possibilities.

The rocketeers tried to attract some of the old Suicide Squad members back, but Apollo Smith turned them down, and Tsien was refused a security clearance because of his nationality. Working for the military had its own complications, and as rumors grew of the Nazis' interest in rockets, all members of the rocket group now had to undergo an FBI security check. The war raging in Europe and the rocketeers' military funding meant that their work had become strictly confidential. "We have to suspect spies under every piece of paper," wrote Malina, half-jokingly.

Within only a couple of months, the project at the Arroyo had suddenly expanded to incorporate twelve full- and part-time workers, including the original rocketeers. A hierarchy was slowly emerging in the previously free-form group. Parsons was named head of the solid-propellant section and was put in charge of preparing JATOs for the upcoming field tests; Forman became head of the machine shop; and Summerfield was named head of the liquid-propellant section. Kármán was assigned the post of director and Malina was the group's chief engineer. The newly enlarged group faced a setback when floods hit the West Coast during the winter of 1940 and turned the usually dry bed of the Arroyo into a thundering river. Boulders crashed by the unprotected shacks, the torrential rain leaked into the explosive powder, and flooding prevented tests from taking place. "This isn't the type of problem that can be solved on order" wrote

Malina. Unfortunately, by the summer of 1941, this was exactly what the group had to do.

Although the rocket work had spectacularly revived itself after its period of stagnation, Parsons' interest in the OTO had not been lessened by his new workload. Indeed, it had grown stronger as he immersed himself in the writings and philosophy of Aleister Crowley. He was initially rather coy about admitting his burgeoning fascination with the occult, especially his encounters with the OTO, to his fellow rocketeers. Nevertheless, the fact that he and Frank Malina had spent "an awful lot of time together" over the past four years meant that both knew of each other's philosophical proclivities. While Parsons was reading Crowley's *Book of the Law* and lectures on yoga lent to him by Wilfred Smith, Malina could be found poring over Darwin's *The Origin of Species* and *The Descent of Man*. "I was personally never sympathetic towards this interest of his," Malina remembered; "I was suspicious of mystics in general ... I used to josh him a bit when he would pull out some of his exotic books, which I just couldn't take seriously."

Many years later Malina recalled Parsons' growing obsession with magic: "He wasn't schizophrenic, but he had two domains which he loved; one was rocketry, where the dream was tangible—where the magic was not resorted to. Then he had a second compartment in his mind where magic fascinated him ... As far as I can remember talking to him about calculations on rocket design, there was no input from what you might say alchemy or magic. In other words he functioned in compartments."

Yet if the two worlds did not overlap, they did share many of the same features in Parsons' mind. Both magic and rocketry had a basis in the imagination and in scientific method. What's more, both promised to satisfy Parsons' desire to escape the earth spiritually and physically. As he wrote to Helen in 1943, rocketry "may not be my True Will, but it's one hell of a powerful drive. With Thelema as my goal, and the stars my destination and my home, I have set my eyes on high."

Though Malina was skeptical, Parsons was quick to invite other less doubtful friends to the house on Winona Boulevard. Perhaps in a final attempt to involve his good friend in his new enthusiasm, one of those he tried to interest was Malina's new wife. With Malina increasingly away from Pasadena, delivering updates on the rocket project to the army air corps on the East Coast, his wife Liljan, eighteen years old and an art student, was often left alone in Pasadena. The Parsons frequently invited Liljan over for dinner to keep her company. "[Parsons] was a nice looking man," she remembered. "Kind of pompous though. He wore a regular suit with a vest, and I don't think I ever saw him without the vest. He liked good music, he liked poetry, he was very well read." On one particular night that her husband was away, she received a call from Jack inviting her to a party at Winona Boulevard.

She recalled: "It was a huge wooden house, a big, big thing, full of people. Some of them had masks on, some had costumes on, women were weirdly dressed. It was like walking into a Fellini movie. Women were walking around in diaphanous togas and weird make-up, some dressed up like animals, like a costume party." Soon she was watching a performance of the mass. "One of these women was rather short and dumpy, and she came out of the coffin and began to flit around in a dance, and I began to think, 'this is really weird, really strange' ... Then some young man appeared in a loincloth and I thought 'This is just too much ... I have to tell Frank about this.'" When she did, Malina simply rolled his eyes and told her not to worry about it. "Jack is into all kinds of things," he said. It was the first and last time Liljan went to one of the OTO parties.

Parsons was also making occasional visits to the recently renamed Los Angeles Science Fiction Society (LASFS). Science fiction was becoming increasingly mainstream. There were now around twenty magazines in print devoted to science fiction—including *Comet Stories, Captain Future,* and *Dynamic Science Stories*—and the quality of writing, which had previously been hit or miss, was improving dramatically. Indeed, John W. Campbell, who had just taken the helm of the magazine *Astounding Science Fiction,* was about to guide science fiction into what many would call its Golden Age. When Parsons heard that one of his favorite writers, Jack Williamson, was due to visit the Los Angeles science fiction group, he wrote Williamson a letter, mentioning that he had been inspired by the rocket design in the writer's story, "Crucible of Power." Williamson, intrigued by Parsons' scientific credentials, agreed to meet with him.

Parsons had a particular interest in one of Williamson's stories that had recently appeared in the fantasy magazine *Unknown.* Called "Darker Than You Think," it presented itself as a found manuscript that told the tale of a newspaper reporter named Will Barbee. Barbee finds himself attracted to a mysterious girl with flame-red hair who calls herself a witch. It slowly emerges that this girl is part of an ancient underground cult of witches that is "awaiting the appearance of an expected leader" known as "the Black Messiah." The witches have long been kept secret from mankind by a vast, hidden conspiracy: "The witches in university laboratories can prove there are no witches. The witches who publish papers can make a fool of one who says there are." Barbee begins to have "dreams" of transforming himself into a great beast and being ridden through the countryside by his red-haired seductress, and he slowly becomes aware that he himself is a witch—in fact he might well be the Black Messiah. Eventually accepting his role, he saves the witches from discovery and eagerly joins the cult. The story's description of a scarlet-haired woman riding a great beast recalled Crowley's own personal mythology, and the tale of Will Barbee seems to have captured Parsons' imagination because it resonated with his own awakening fervor for the OTO.

Jack Williamson had majored in chemistry before taking up writing full-time. Lean, bespectacled, slightly hunched over, he was, at the age of thirty-two, the elder statesman of the genre. He found Parsons "an odd enigma" and was fascinated by his "unexpected" interest in occult ideas. "I don't remember much talk about the occult. He was I think rather reserved about his interest in it, at least with skeptics." Williamson had written "Darker Than You Think" as a study of his own inner conflicts as he grappled with them under psychoanalysis, and he hadn't the slightest interest in real-life cults. Nevertheless, there was something about Parsons that intrigued him: The occultist-rocket scientist could have been a character from one of his own stories. Following their meeting, Parsons invited Williamson and another science fiction writer, Cleve Cartmill, to pay a visit to the house on Winona Boulevard and attend the mass. The two writers were underwhelmed by their visit. "The ritual was disappointingly tame. There was no nude virgin on the altar. Satan was not invoked," remembered Williamson.

One LASFS member who was attracted to the ceremony for more than laughs or research purposes was Grady McMurtry, a senior at Pasadena City College. He had become somewhat enamored of Parsons through his appearances at the LASFS meetings at Clifton's Cafeteria. Parsons was older, better read, and had an unquestionable charisma, and McMurtry remembered being particularly impressed by the "fancy Russian cigarettes" Parsons smoked, little knowing that a few months earlier all the rocketeer could afford were scrounged dog-ends.

Parsons proceeded to act as a mentor of sorts to the young McMurtry, introducing him to ancient mythology and Frazer's *The Golden Bough*. McMurtry wrote poetry, and Parsons, whose own verse output was increasing as he read more of Crowley's work, offered to lend him some guidance. Soon McMurtry was being invited back to Parsons' house to listen to Parsons' favorite music, Stravinsky's *The Rite of Spring,* and for post-LASFS drinking sessions alongside the visiting Jack Williamson and Ed Forman. "Must have drank a quart and a half of beer," wrote McMurtry after one particularly heavy session; "talked about rockets, witchcraft, etc." Betty Northrup, Helen's vivacious sixteen-year-old half-sister, also joined in the discussions. She had recently moved in with Parsons and Helen while she finished high school, presumably in an effort to escape her father. She got along well with Parsons; the two would discuss various ideas of books they could write. The pair would often go on long walks in the nearby mountains along with Helen and McMurtry, and Parsons would smoke his pipe and talk of nothing but poetry.

On February 15, 1941, Jack and Helen Parsons were finally initiated into the Agape Lodge of the OTO. Parsons took as his motto, "Thelema Obtentum Procedero Amoris Nuptiae." In a testament to his less-than-perfect school days, it was horribly incorrect Latin, but the general meaning seemed to be, "The establishment of Thelema through the rituals of love." Transliterated into the Cabbala, the initials of his motto gave him the magical number of 210, with which Parsons would take to signing all his official OTO correspondence.

A delighted Wilfred Smith wrote Crowley, "I think I have at last a really excellent man, John Parsons. He has an excellent mind and much better intellect than myself ... JP is going to be very valuable." Parsons' value was not least to Smith himself, for his initiation came at a particularly testing time for the head of Agape Lodge. One of the OTO's members, the jeweler and astrologer Max Schneider, strongly disapproved of Smith and was jealous of his position as the chief of the OTO in California. He wrote letters to Crowley claiming that Smith was abusing his position by prostituting some of the OTO's members, and he made the far more serious accusation that Smith was embezzling money destined for Crowley himself. Parsons was soon to discover that bad-mouthing fellow OTO members was an addictive habit for the inhabitants of Agape Lodge. It was a habit that Crowley also enjoyed, playing his disciples off against one another in order to test their loyalty. Schneider's charges appear to have been groundless, but Smith was grateful that he could trumpet an accomplishment: Parsons could not only increase membership of the Lodge but also get Smith back into Crowley's favor. Other members of the OTO thought Parsons might be a blessing for the stagnant order. Jane Wolfe wrote excitedly to Crowley's second-in-command, Karl Germer, an expatriate German in New York who as treasurer was in charge of collecting the OTO's donations and fees. She had pinned her hopes on the new initiate, writing, "This year there is more hope, as we have an A1 man joined up, Crowleyesque in attainment as a matter of fact."

From the start, Parsons displayed a great desire to involve himself with the OTO. He ambitiously declared that he would begin a course of talks at his house in Pasadena in order to enlarge the group's membership and to find new "prospects" among his friends. Soon he persuaded Helen's half-sister Betty and his science fiction friend Grady McMurtry to join. But his fervent planning and boosting for the OTO meant that he was not seeing as much of the rocket group socially as he had before. Noticeably he was absent from the parties Kármán and his sister Pipö gave at their house. When it was remarked to Kármán that Parsons hadn't been attending because he had become involved in a strange cult, Kármán simply laughed as Malina had. "Oh Jack," he said, amused. "He's just a little crazy, don't pay attention to him."

It had taken almost two years from his first visit to the mass for Parsons to become a fully paid-up member of the OTO. The long hours spent working on the rockets, Helen's initial reluctance to join, and Parsons' own ambivalence regarding Wilfred Smith had all helped delay the process. However, by early 1941 these obstacles had been surmounted: The rocket work was fully funded, Helen had slowly become fascinated by Crowley's religion ("We met something that was right!" she would recall), and Parsons' hesitancy over Smith had all but disappeared. Indeed, Parsons began to see Smith and the OTO as more than just his guides through the arcane realms of magick. They seemed to offer him something that even the Suicide Squad could not. In a letter to Smith, Parsons made a poignant admission: "You know, I was an only and lonely child, and it is a fine thing to inherit such a large and splendid family. I never knew a father, and it is nice to have one now."

8

· · · · · · · · · · · ·

Zenith

What did you go out into the wilderness to look at?
A reed shaken by the wind?

—Matthew 11:7

The desert has always been a testing ground. It is a place where devils tempt morals, heat sears endurance, and sheer soundless magnitude confounds the sanity of the solitary man. But it is also fit for transcendence and communion: "There are places where humours and fluids become rarefied," says Jean Baudrillard, "where the air is so pure that the influence of the stars descends direct from the constellations." For Aleister Crowley the "beloved Sahara" embodied just such a spirit: elemental, ancient, silent, it was a stage facing out onto the universe on which he could act out his rituals and incantations.

But in 1941 the Californian desert seemed very different to the ancient deserts of Islam in which Crowley had dwelt. While paintings and carvings of antique civilizations peppered the caves and hollows of the Mojave, the rest of the desert was moving to a more fractious, modern beat. Airfields and explosives factories, cement mills and borax blowers, bombing ranges and oil fields now inhabited the landscape. The desert had been turned into science's giant laboratory, an almighty backyard in which scorched earth and noise exploded out against the fossil silence.

Twenty years before, Robert Goddard had been chased into exile in the desert—as all good prophets are. For Parsons, however, the desert had always represented a vast playground. The wide open spaces where he could fire rockets into the eye-blue sky had long offered him a welcome respite from the slide rule and graph paper. Indeed, the Suicide Squad often traveled en masse into the desert to unwind. Hunting trips were the standard entertainment. Armed with rifles and shotguns, the men would take shots at any cottontails and jackrabbits that strayed across their path—and sometimes at each other. It was on trips such as these that Parsons and Forman were famed for enacting duels. "Jack and Ed Forman decided to play a game," remembered Caltech graduate student, Homer J. Stewart. "[They stood] fifty yards away from each other, and they shot at each other. Now the rule was that who came closest without hitting won."

The duels attested to the pair's bare-chested machismo and the continuing closeness of their relationship. "They were both daredevils," remembered Jeanne Ottinger, Forman's stepdaughter; "whatever Jack suggested, Ed was going to go right along with." But the duels were also declarations of intent; both men reveled in testing the boundaries of the sane and the safe, just as they did in their work on rockets. "They both wanted to try anything, do anything," remembered Ottinger. "They thought that nothing was impossible." Neither man was ever hurt in the duels; indeed, the game usually ended when dirt flew up inches from one of the men's feet. For the other members of the Suicide Squad (how ominous the name must have sounded during these moments), the game was entirely characteristic of the two—somewhat discomfiting, somewhat rash. But they took it, as they did so many eccentricities within their ranks, with a shake of the head and an exhalation of breath. "That was Jack, that was Ed—they were crazy, what could you do?"

The desert would now be the backdrop for a project whose future hung uncertainly between the possible and improbable. In the spring of 1941, the Suicide Squad informed the army air corps that the group's JATO rockets would be ready for testing in early August. Parsons had been changing and improving the JATO's fuel. Nevertheless, it still seemed to be made up of the contents of a schoolboy's desk. He pressed amide black powder mixed with corn starch and ammonium nitrate into the blotting paper-lined rocket, the whole mixture being bound together with Le Page's all-purpose stationery glue. Fred Miller, an engineer from Caltech who had recently joined the expanding project, called it "the goop," but officially it bore the more specialized name of "GALCIT 27," since it was the twenty-seventh unique fuel Parsons had created in the last few months.

The rocket motor which held this powder seemed equally primitive—one foot long, three inches in diameter, and made of one-inch-thick steel, it held three pounds of propellant compressed under 1,800 pounds of pressure. The three pounds of fuel would burn for a wonderfully lengthy twelve seconds while delivering a small but punchy twenty-eight pounds of thrust. The propellent was by no means as powerful as some of the liquid rocket fuels they had been experimenting with, but it was enough to aid a small plane in getting off the ground and thus provide that most important commodity for the rocketeers—a spectacle.

The group decided to adapt a small 753-pound single-engine Ercoupe, a civilian monoplane, for the tests. The plane was easy to handle—anyone who could drive a car could fly it—it didn't go into spins, and it was very cheap in case anything went wrong. One of Kármán's graduate students, the thirty-two-year-old Homer Boushey, a tall, bluff San Franciscan who was also a lieutenant in the army air corps, agreed to be the pilot.

Boushey had also bought in to the romance of rocketry. From his earliest days in flight school, he had been fascinated by Goddard's work. Displaying the precociousness of almost all rocket amateurs of the time, he had even corresponded with Goddard about the possibility of developing rocket propellers. Such an offbeat hobby had not been looked on kindly by his army superiors, however, and they had ordered him to forget rocketry and instead to concentrate on "flying the mail" in his open-cockpit biplane. Their skepticism, however, failed to quench his interest, and now when offered the chance to take a leading role in a serious rocketry experiment, Boushey found it hard to resist.

He almost immediately began to wonder if he would regret this decision. A week before the flight tests were to begin, Boushey visited the rocketeers to see how the JATO rockets were coming along. What he saw could not have been encouraging. The rocketeers were casually strolling to and fro in various states of undress. Martin Summerfield had even set up home in one of the fifteen-foot-by-fifteen-foot tin shacks that the Suicide Squad had built; his underpants and socks hung unceremoniously from the gauges and motors.

Parsons himself was bedraggled and worn out. He had, since April, been holding the talks he had promised the OTO. These took the form of twice weekly discussion groups held at his home that touched on literature and mysticism. Members of the OTO were present, as well as a few students from Caltech who had become intrigued by Parsons; his interests seemed a world away from the campus pastimes of glee clubs and golf. The evenings dragged on late as records were played and poetry was read and discussed. At the same time, the OTO house at Winona Boulevard had undertaken a new publicity drive, holding parties to raise money and spread awareness of the cause. Parsons was at the center of these, joking and declaiming, his eyes wandering over the young bohemians who had been curious enough to attend. Spiked punch would be served before Parsons diligently and persuasively suggested that if his guests wanted an "experi-

The *Los Angeles Times* leads with the news of Parsons' death. *Courtesy of the* Los Angeles Times

Investigating the explosion that killed Jack Parsons. Note the washtub on the right. *Department of Special Collections, Charles E. Young Research Library, UCLA*

Parsons at the age of eighteen. He would be a lifelong fencing enthusiast.
Warburg Institute

Amazing Stories #1, "the magazine of scientifiction."
Reprinted by permission of the agent for the estate of Frank R. Paul

The Suicide Squad rests from its labors. From left: Rudolph Schott, a student assistant; Apollo Smith; Frank Malina; Ed Forman; Jack Parsons. *Courtesy NASA/JPL-Caltech*

The Los Angeles Science Fantasy Society convenes at Clifton's
Cafeteria. A young Ray Bradbury sits two from the left. Forrest
Ackerman, the "Number One Fan," sits to his left. Robert Heinlein,
with mustache and glasses, sits five from the right. Jack Williamson
stands six from the right. *Los Angeles Science Fantasy Society, Inc.*

"The Wickedest Man in the World," Aleister Crowley.
Warburg Institute

The Ordo Templi Orientis performs the Gnostic Mass in February 1939. From left, Wilfred T. Smith (high priest), Regina Kahl (high priestess), Luther Carroll (deacon).
Warburg Institute

Theodore von Kármán (in black suit) at the Ercoupe tests jots down some equations for the camera. He is surrounded by, from left, Clark Millikan, Martin Summerfield, Frank Malina, and Captain Homer Boushey. Note that Parsons is cropped out on the left. *Courtesy NASA/JPL-Caltech*

Parsons, Forman, and Malina pose proudly in front of the Ercoupe. *Courtesy NASA/JPL-Caltech*

The first Aerojet office. Parsons stands to the right side of the doorway, Andy Haley to his left. *Aerojet: The Creative Company/Carson Hawk*

Parsons visits the exiled Wilfred Smith and Helen at the turkey farm in the desert. *Warburg Institute*

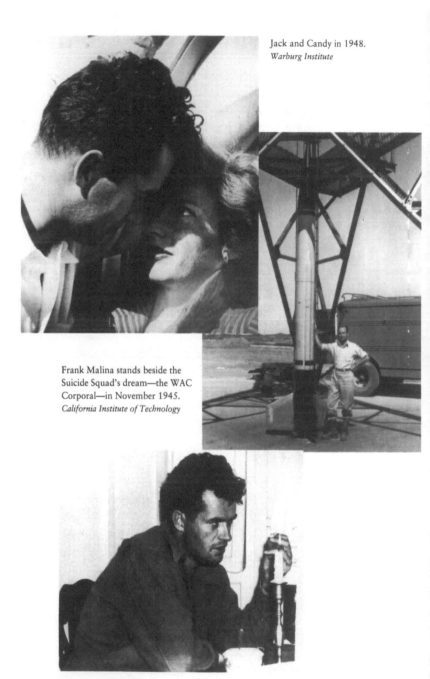

Jack and Candy in 1948.
Warburg Institute

Frank Malina stands beside the
Suicide Squad's dream—the WAC
Corporal—in November 1945.
California Institute of Technology

Parsons in the late 1940s, shortly before his death.
Warburg Institute

ence," they really ought to watch the Gnostic Mass in the house's attic. In the mornings after his late-night activities, an unkempt Parsons could be seen shuffling across the Arroyo, flinching in perceptible agony as his rockets belched and screamed.

On the day that Boushey visited, the rocketeers were running a technical rehearsal. Parsons had set up six JATOs on a metal rack, simulating the rack that would be attached under each of the test plane's wings. The group of rocketeers, including Kármán, waited anxiously fifty yards away, behind a wall of sandbags. Parsons, cigarette in hand and feeling poorly, rushed through the countdown in his usual style. He pushed the firing button. "Explosions occurred like on the Fourth of July," remembered Kármán; "nozzles flew off in all directions. It couldn't have been any worse." Parsons was left horribly sobered.

If anyone was more perturbed than Parsons, it was Boushey. After all, in seven days Boushey would be sitting about six feet from twelve such rockets in a plane so fragile that it could be torn apart if all the rockets misfired at once. Although the rocket exhaust nozzles were attached to the motors with shear bolts—designed to break in half under extreme tension and thus relieve pressure in the motor before a catastrophic explosion could take place—this could not prevent rocket shrapnel from hitting and piercing the fuselage.

For the majority of the tests at the Arroyo, the powder had worked just fine. Why was it suddenly exploding? The tests scheduled for the next week could not be delayed, so Parsons furiously analyzed, purified, and retested his fuel mixture, anxiously trying to find out what was at fault. Writing to his parents, Malina summed up the thoughts of all the rocketeers. "The phase of research that was supposed to be solved has been cutting up. At the end of today I felt that nothing short of a miracle would pull it through by the date we are supposed to meet." On August 6, the scheduled start of the tests, a miracle had yet to materialize. Nevertheless, the group knew a delay would jeopardize their funding, so it headed to March Field Air Corps Base, the oldest air base in the West. Located in Moreno Valley, sixty-six miles east of Los Angeles, March Field's runways and hangars had been built on the remnants of an old cattle ranch. The original residents of Moreno Valley had moved there in the early 1880s, but after a catastrophic loss of water rights was followed by a severe drought, they had no choice but to leave. The more expensive homes were carried away whole on the trailers of steam-powered tractors, while ingenious thieves relocated those houses that had been left behind to unknown locations. March Field had replaced them and now stood as a lone bastion of human life in the midst of an encroaching desert.

A fresh batch of eighteen JATOs had been made that morning and driven to March Field. The rocketeers marked the runway in fifty-foot intervals in order to measure the takeoff distance of the Ercoupe. Observers on the ground clasped stopwatches with which to measure takeoff time. Initial tests were carried out to measure the performance of the Ercoupe without the JATOs. Then a static test was performed in which the Ercoupe was anchored to the ground with one rocket placed under each wing. The rockets fired perfectly, the flames easily clearing the fuselage of the plane, leaving it undamaged.

By the eighteenth test of the day, the heat was rising and the inevitable had to be faced: It was time to test the JATOs in flight and find out how much extra speed the rockets could offer to the plane. To be on the safe side, the rocketeers installed only one jet unit under each of the two wings. Boushey lifted the tiny silver-and-red plane off the runway and flew to 3,000 feet. Straining their eyes against the desert sun, the rocketeers could barely see the plane as it was swallowed up in the dark midday blue of the sky. Boushey made a couple of passes over the field, then ignited the rockets. Suddenly there was no doubt as to where the plane was. Two brilliant white jet trails appeared in the sky behind it, seemingly attached to the Ercoupe like cords on a kite. The trails increased in length and the rocketeers watched, mouths agape. The sight was, after all, in the days before jet trails became commonplace in the world's skies. All seemed serene until one of the trails abruptly cut off. All eyes anxiously scanned the remaining jet trail, which was now itself dissolving in the wind. Slowly emerging out of the blue, the Ercoupe could be seen hurrying down towards the runway. When the plane touched down, it was clear to all what had happened—one of the rockets had exploded. Fortunately, it had done so near the end of the run, and the safety devices had worked; the shattered apparatus had cleared all parts of the plane and fallen harmlessly to earth. Without further ado, and with some relief, tests were wound up for the day. It appeared the rocketeers had escaped a serious mishap by the skin of their teeth—and thanks to the skillful flying of Boushey.

Returning to Pasadena for the night, Parsons continued tests on the fuel, but two days later he had come no closer to solving the problem of their rockets' volatility. Nevertheless, the rocketeers kept to their strict timetable and reconvened on August 8 at March Field to begin an additional series of grounded tests. Test Number 20 called for another static firing in which the Ercoupe was anchored to ground, though this time the team installed four JATO rockets, two under each wing. Boushey fired the ignition and almost immediately an explosion occurred. The rocketeers could only stand back and watch as the faulty rocket shot sparks and flame out of an increasingly dense haze until finally the Ercoupe was engulfed in smoke. As the motor spluttered to a halt and the light desert wind swept the white shroud away, the airplane revealed was older and more battered than the one that had stood there ten seconds before. The rocket rack had been pulled loose, the wing covering was wrenched out of shape, rivets were scattered about the runway, and the white-hot nozzle of the rocket motor had ripped a ten-inch hole in the tail end of the fuselage. Boushey clambered out, slightly shaken. The team moved in to inspect the damage. "Well at least it isn't a big hole," piped up one voice. It was met with a chastening silence. Testing was postponed until repairs could be made.

Back at the Arroyo, Parsons finally realized what was happening to his fuel. Out of all the tests at March Field, those conducted with JATOs that had just been pressed into their chambers—JATOs that were fresh—worked perfectly. However, those motors made more than twenty-four hours before the test had exploded. Parsons could see the problem now. If left to sit, the powder in his fuel would contract in the cold of the night and expand in the warmth of the day. This would cause long fissures and cracks to appear in it. The result was that rather than burning evenly and slowly, the fuel ignited in an instant, flame spreading rapidly through the gaps and spaces. Pressure inside the motor would increase rapidly and the rocket would be turned into little more than a bomb. If Parsons wanted his rockets to work, he would have to press the powder into the chamber just before the test.

His breakthrough came just in time, for never was success more important than on the next day, August 12. The experiment the armed forces were most eager to view was scheduled to take place: a takeoff aided by rocket power. Parsons, Forman, and the mechanic Fred Miller worked through the night, preparing eighteen fresh JATOs for use. The loading of the motor with the "goop," one inch at a time under pressure, was painfully laborious, each rocket taking forty-five minutes to fill. Only adrenalin kept them awake. As the sun rose, they loaded a truck with the canisters and raced over to March Field. Every minute was crucial. The longer the delay in getting to March Field, the more likely it became that their tests would fail. As soon as they arrived, Parsons crawled underneath the plane and attached the rockets to its wings in a matter of minutes. A brave Homer Boushey seated himself in the cockpit.

A Super-8 film made of the tests gives a particularly vivid impression of that day. Scratched and jumpy, the color film captures the thin layer of ashen morning mist that lies across the scrub-covered, flat desert ground. The saw-bladed Sierra Madres rise up in the distance. The metallic Ercoupe glints in the morning sun. The film cuts to Parsons, who stands near the plane along with the rest of the Suicide Squad. He is looking intently at the Ercoupe, snatching smoke from his cigarette. In fact, all the team are smoking and talking, talking away their fears, exhaling their worries in fitful plumes of white smoke. The massive quantities of explosives at the test site does not seem to have curbed any of the team's desire to light up. Parsons looks young, or rather his body does. He is still only twenty-six years old, and hunching forward in his short sleeve shirt, his posture is decidedly adolescent. His face ages him, though. His eyes have a knowing look; his jaw is stretched by a smirk; the razor-thin moustache he has grown lends him a mischievous Mephistophelean air.

There is some good-humored posing for the camera, much laughter at unheard jokes. Clark Millikan, head of the GALCIT wind tunnel, whose initial skepticism about rocketry has by now been replaced by an opportunistic enthusiasm, stoops down as if in conversation with the mechanics. He is intent on showing himself to be in the thick of things. In another shot, an official Caltech photographer is taking a picture of Kármán, in black suit and crumpled white fedora, writing equations on the wing of the Ercoupe. Malina, Summerfield, and Boushey are leaning over Kármán's shoulder, as is Clark Millikan. What the film revealingly shows and what the resulting photo—one of the best known of the tests—has left out is Parsons rushing to get into the picture. Upon arriving at the plane, he leans on the wing to Millikan's right, hand on chin, in an exaggerated pose of studiousness. Perhaps because of his lack of reverence, perhaps because he was not officially affiliated with Caltech, or perhaps because he just didn't get there before the shutter closed, in this public relations photo, which is reprinted frequently in many of the staple textbooks on rocketry, Parsons has been cropped from view.

The Super-8 camera captures some twenty to thirty people watching the tests—friends, air force observers, Caltech students. Most prominent among them stands the white-bearded William F. Durand, chairman of the National Advisory Committee on Aeronautics (NACA), editor of the first encyclopedia of aeronautics, the grand old man of United States flight, and an old acquaintance of Kármán. The film jumps and the Ercoupe is suddenly seen speeding down the runway. All four rockets have been ignited and a vast plume of smoke follows the tiny aircraft. The plane is hurled into the air as if from a slingshot, the rockets still firing and smoke still billowing. It climbs sharply at a near fifty-degree angle before the rockets cut out and Boushey levels the plane off, circling the airfield and landing. The film jumps again to the long-limbed Boushey bounding out of the plane, grinning broadly. "None of us had ever seen a plane climb at such a steep angle," commented Kármán. It was the first American airplane ever to fly with rocket power.

The rocketeers were jubilant. For the next test they strapped six JATOs to the Ercoupe and measured the takeoff distance and takeoff time. The results exceeded their highest expectations. Both had been reduced by nearly 50 percent, with the takeoff distance falling from 580 feet to 300 feet and the takeoff time decreasing from 13.1 seconds to 7.5 seconds. The plane's speed, measured at the airplane's maximum ceiling height of 11,400 feet, jumped from 62 to 97 miles per hour, an increase of 56 percent. The effects were exactly what the armed forces had been looking for. These rockets could be attached to heavily laden bombers, allowing them to take off from short, makeshift jungle airstrips. Airplanes which could never before have built up enough speed to take off from the shortened runway of an aircraft carrier would now be able to do so easily.

Over the next two weeks, more tests took place. When Clark Millikan piloted his own plane, a red Porterfield, alongside the Ercoupe to act as an experimental control, he was left in the JATO-powered aircraft's wake. Over the next two weeks,

152 JATOs would be used in 62 tests without any serious explosions, largely thanks to the early-morning delivery dashes to the airfield by an exhausted and bleary-eyed Parsons.

Most of the planned tests had been completed when Kármán, who had a showman's eye for publicity, suggested that the plane try taking off without using its propeller, as a purely rocket-powered aircraft. The propeller was removed and safety posters that read, "Be alert! Don't get hurt!" were pasted over the large gap that now let wind into the cockpit.

The rocketeers attached twelve rockets to the plane so that it now resembled a giant firework. Boushey fired the rockets from a standing start—first six, then another six—but the plane failed to reach takeoff speed under their thrust alone. Ever game, he suggested that a truck tow him to about twenty miles per hour before he fired the rockets. Once again the silent film pays dumb witness to an astonishing sight. Boushey grips a rope through the open cockpit window of the Ercoupe with his left hand. The sun glints off the watch he is still wearing on his wrist. The other end of the rope is tied to the back of a truck that is driving down the runway. The plane, its nose swathed in the safety posters, trundles along behind it, gathering speed, up to the point, as Boushey said, "somewhat below that wherein the arm-bone would have been pulled out of the shoulder socket." Letting go of the rope and pulling his arm back into the cockpit, he flicks the ignition on all twelve rockets at once. The plane leaps into the air with a roar, shooting high on a bridge of smoke. Then, as the rockets cut out, it plunges back down and slams into the runway, hard but safe. A bruised Boushey clambers a little more tentatively out of the cockpit than before. That would be all for now.

The first flight of an airplane powered solely by rockets had been brief, but it was a signal of things to come. That year the NAS not only renewed the grant to the rocketeers but increased it to $125,000. There would be no more Buck Rogers jokes. Boushey, for his part, went back to rejoin his squadron, where he could be heard shaking his head and muttering about "those birds who got me to fly an airplane without a prop."

Amidst the chorus of celebration sounded a few dissonant notes. Ever since the military had stepped in as the Suicide Squad's main benefactor, levels of security had soared and a restrictive air of secrecy had overtaken the project. When Frank Malina wrote to his parents about the Ercoupe tests, he could only hint obliquely as to the rocketeers' work: "We have had success this week with our rocket project that exceeded even our highest expectation. Wish I could tell you more. We have had a taste of what we have been striving for the past three years." The classified nature of their project raised even more pernicious problems. At an aerodynamics department party, Frank Malina was happily talking of the recent success that had transformed him from a graduate student with a harebrained idea to a serious pioneer in the field when he found himself confronted by Clark Millikan, his old bête noire. With undisguised glee, Millikan informed Malina that he had been told that Malina, Sidney Weinbaum, and two or three other members of the squad belonged to the Communist Party. The FBI had apparently provided Caltech with this information following their security checks into the rocketeers. Indeed, Parsons' name had come up, too, and he had been questioned by the FBI about the Communist group's activities. Being a member of the Communist Party was not illegal at the time, but considering the sensitive military work that the project had become involved in, it was not comforting news to Malina to hear that the FBI knew of it. If any of the rocketeers lost their security clearance, their ability to stay at the forefront of their field would be jeopardized.

Just as ominous were the possible repercussions from Parsons' recreational activities. One morning in November 1941, Malina received a phone call telling him that a GALCIT night watchman was in jail. His name was Paul Seckler, and along with his wife Phyllis he was a resident of Winona Boulevard and a member of the OTO (Parsons had gotten him the watchman job). The previous night Seckler had gone to Parsons' house for one of his twice weekly meetings. Ed Forman had also been there. But something had happened on this night that had deeply affected Seckler. Malina thought a séance might have taken place. More likely, drink, drugs, and perhaps a magick ritual were involved. All that is certain is that Seckler, who was known in the OTO to suffer from fits of drunken violence, ran out of Parsons' house, half-crazed, clutching a pistol belonging to Parsons. He found a car parked on a nearby street in which a young couple was engaged in a romantic entanglement and forced the pair out at gunpoint. He then took the car and drove around Hollywood for a few hours. Calmed somewhat by his flight, he drove back to Pasadena in the early hours of the morning and found the police waiting for him at the Colorado Street Bridge.

Malina was shocked. With the tightened security procedures at the rocket project, secrecy was paramount. Now here was one of his staff getting arrested for hijacking a car! Malina immediately drove to the jail to talk to the chief of police, who agreed to keep the story out of the newspapers. However, when he spoke to Seckler, Seckler was "very vague" and refused to talk about the incident. When Malina then called on Parsons and Forman, they, too, wouldn't say a word. The incident signaled a loosening of the close bonds that had tied Malina and Parsons together. The Suicide Squad was growing rapidly, and each rocketeer had become involved not only in his own scientific problems but also in his own social world. Malina recalled, "Our relationship had become much more tenuous, so I didn't really know all that was going on." Seckler was sent to San Quentin State Prison for two years, leaving Malina to wonder whether Parsons' extracurricular enthusiasms might be getting out of hand.

On December 7, 1941, Pearl Harbor was bombed, prompting the United States' formal declaration of war on Japan. America's industrial assembly lines began working around the clock in the rush to war. Typewriter factories began making machine guns; Chrysler started building tanks. Scientists, military officers, economists, executives, and public officials were all brought together to pool their talents in order to find solutions to urgent wartime problems. The result was an economic boom which the United States somewhat guiltily, embraced. The country's pre–Wall Street crash confidence had returned.

Exempted from the draft because of their military work, the rocketeers now knew they must focus all of their attention on getting results and getting them quickly. Finding a stable liquid fuel was now their main focus, for liquid fuels were more powerful than solid fuels and in addition they could be turned on and off at will by regulating the flow of fuel into the rocket motor. Once you lit a canister of solid fuel, its combustion could not be stopped. However, as the Suicide Squad's first explosive experiments in 1936 had proved, developing liquid fuels was no simple proposition. None of them currently known could be stored safely, especially not in the rough wartime conditions that the navy had stipulated.

Martin Summerfield had been working on the liquid side, following Parsons' experimental lead, testing several combinations which used gasoline as a fuel and red fuming nitric acid (RFNA) as the oxidizer. His trials, conducted at freshly dug test pits in the Arroyo, had met with predictably varied results. The first two motors had both failed unspectacularly. When the third motor was tested, it exploded with such ferocity that it set fire to the sides of the test shack. The clamps holding the motor in place were bent out of shape, gauges broke, and the rocketeers were sent scrambling for cover as the wall of sheet metal protecting them buckled.

After the flames had been doused, Parsons and Summerfield conducted a postmortem on the battered and burst apparatus. They concluded that the main cause of the explosion was a throbbing action in the motor. Once the fuel had been ignited, the motor sporadically began to pulse, its sides expanding and contracting while the thrust fluctuated wildly. The throbbing was slight at first but increased in intensity until at the fourth or fifth pulsation the motor blew up. They thought the gasoline fuel was the cause of the throbbing, and they tried differing amounts of fuel and oxidizer and different pressures in the chamber to no avail. They grew all the more concerned when, in the wake of Pearl Harbor, the army informed them that it expected to view flight tests of the liquid-fueled motors in the spring of 1942. This time the test plane was to be no lightweight Ercoupe but a 14,000-pound Douglas bimotor bomber, the A-20A. It was time for Malina to come up with a solution.

During a visit to the Naval Engineering Department Station at Annapolis, Maryland, in February 1942, Malina learned of a liquid named aniline. It was a hypergolic chemical, meaning that it ignited spontaneously when added to the nitric acid the rocketeers were using as an oxidizer, and Malina initially thought it could replace the troublesome ignition mechanism they had been working on. Going one step further, however, he telegrammed Summerfield and told him to get rid of gasoline altogether and replace it with aniline. By the time he returned to Pasadena, the rocketeers' motor was running smoothly; all throbbing had been eliminated; and the new fuel's unique property of spontaneous ignition had allowed the group to dispense with the ignition system, greatly simplifying the motor's design. Of course, aniline had its own share of problems. It was highly toxic and could rather worryingly be absorbed through the skin. When tests took place, prevailing breezes were checked and the immediate vicinity was evacuated. Nevertheless, with the test date approaching, this new fuel was a godsend. The rockets fired with an awesome, controlled power. Malina wrote an ecstatic letter home: "We now have something that really works and we should be able to help give the Fascists hell!"

The first liquid JATO tests were held at Muroc Auxiliary Air Field in the heart of the Mojave Desert, across the mountains from March Field. This unforgiving landscape contained no traces of civilization. The air field is situated on the eastern shore of Muroc Dry Lake, a barren, hard-packed playa—a perfectly flat and saucerlike depression that was already being used for automobile speed tests and as a bombing range. On April 15, 1942, the team laid out a stripe three feet wide and twelve thousand feet long on the lake bed as a guide for the pilot, Major Paul Dane, another of Kármán's former students. In contrast with the mood at the Ercoupe tests, spirits were high. Parsons was still concerned with his solid-fuel rockets, and he only made the one visit to Muroc, missing many of the actual test flights. Nevertheless, his playfulness seemed to have infected the entire group's preparations. Even the most senior members of the squad were not immune to the pranks and slipups which so often seemed to accompany the rocketeers' experiments.

When a navy observer flew to Muroc in the latest state-of-the-art fighter plane, he asked Kármán, who was also paying a visit to the test field, if he would like to view it. The rocketeers wondered aloud whether this was a wise thing to do—Kármán had an uncanny knack for wreaking havoc on experimental equipment. Laboratory apparatus would often lie in ruins following Kármán's enthusiastic prods and pokes. Paying no heed to these warnings, the pilot motioned for Kármán to climb into the cockpit and began to explain the array of dials and switches in front of him as the diminutive Hungarian gazed on excitedly. Suddenly Kármán pointed to a handle by his foot and, asking what it was for, gave it a sharp pull. The navy pilot's horrified cry, "No!" was drowned out by a large crash behind the wing. The rocketeers scurried away from the airplane, expecting an immediate detonation. Instead, two giant balloons began to inflate under each wing; Kármán had released the water flotation gear. He swiveled in his seat delightedly watching the balloons expand, while the Navy observer clutched his head in despair. "My Lord," he gasped. "How will I explain this at my base— flotation gear activated in the desert!" The rocketeers allowed themselves a knowing smirk.

The hulking A-20A bomber that the group planned to use for their test had already been fitted with two liquid-fuel thrust rockets. With the fuel and oxidizer tanks built into the wings, each motor was capable of delivering a thousand pounds of thrust for twenty-five seconds. They were fired by Beverly "Bud" Forman, a cousin of Ed Forman's, sitting in the tail gunner's seat.

After several static tests, it was time to attempt a rocket-powered takeoff. The engines were revved up and the cumbersome black bomber lumbered down the runway. When the rockets ignited, the plane leapt forward, starting like a bull to which a white hot brand had been applied. Inside the plane, Dane and Forman could feel the heat of the boosters on their faces, clean through the aluminum of the plane's body. In a cloud of swirling acrid smoke, the plane charged into the air. The team recorded the point of take-off and made a series of quick calculations. The two liquid-propellant rockets had chopped one-third off the propeller plane's normal takeoff time and distance. The rocketeers on the ground cheered as Dane indulged in some acrobatic celebration.

The JATOs had worked. After years of trial-and-error experimentation and painstakingly thought-out theory, the team had reached their goal. Just a few weeks later, the squad had signed their first production contract, agreeing to sell sixty Jet-Assisted Take-Off engines to the army air corps. The Suicide Squad, now officially GALCIT Project Number One, had invented a product that was unique, successful, and, more importantly, marketable. It wasn't time to sell out, but it was certainly time to cash in.

Man cannot live on gunpowder, smoke, and adrenalin alone. After years of privation and nonstop work, the founding members of the Suicide Squad felt that it was time to reap the benefits of their achievements. Indeed, for Parsons and Forman, still unaffiliated to any institution of learning, even the recent successes did not entirely ease their anxiety. If the JATO projects were cancelled, they would be unceremoniously returned to the same place they had been in 1936, mixing explosives for powder companies in the desert. They made their displeasure known and Malina for once agreed. "I shared the opinions of Parsons and Forman that after the efforts we had made during the previous five years we should participate in the exploitation of our ideas."

Even before the liquid-fuel rockets were tested, Malina had approached Kármán with the idea of forming a company to produce the solid fuel JATOs that the rocketeers had invented and sell them to the army. Such a project would not be easy. Despite the commercial uses made of the wind tunnel, Caltech was primarily an institution for education and research, and the administration had made it clear that while it now condoned the research work being done by the GALCIT group, it had neither the time nor the inclination to immerse itself in large-scale engineering development and production. Old man Robert Millikan, still the president and moral center of Caltech, lowered his eyebrows and shook his head at the idea of Kármán and Malina—supposedly men of pure science—turning their attention toward that dread brother of learning, industry. Millikan's concerns were ironic, considering the close ties the coming war would foster between industry and science, but it was not just academia that disapproved. Further objections were raised by General "Hap" Arnold of the army air corps, who returned Robert Millikan's compliment by saying that he would never trust an academic institution to go into manufacturing in any case.

In order to avoid angering either of the rocketeers' main benefactors, Kármán suggested that it might be best to approach an existing aeronautical company about producing JATOs. The rocketeers could then lend their expertise to the work and participate in the profit sharing, without upsetting Caltech or worrying the military. He contacted all the aircraft firms in Los Angeles to see which would like to share in their endeavour and set up a rocket engine division making JATOs. Since the bombing of Pearl Harbor, all six of the large aviation firms located in the Los Angeles area—Douglas, Lockheed, Vega, Northrop, North American, Convair—had accelerated into a triple-shift, around-the-clock schedule. President Roosevelt had called for the construction of 60,000 planes a year—more than had been manufactured in the United States since the invention of flight—and the air companies now employed a workforce of over 100,000 people, up from 13,000 in 1939. Even Parsons' mother Ruth was now working six days a week at Lockheed in order to support herself. But despite the industry boom, Kármán's high standing within the aeronautical world, and the successes of the Ercoupe and A-20A tests, not one company would take Kármán up on his offer. Outside of Caltech the stigma of rocket propulsion was still strong, despite the army's interest, and was shared by the pinstripe as much as the gown. The group would have to go it alone once more.

Well, not quite alone. Old man Millikan and General Arnold had a point Kármán could see all too well: Scientists rarely make good businessmen. It is little surprise that Frank Malina, the only member of the Suicide Squad to make a significant amount of money in his later years, did so partly by accident. Realizing his own inadequacies, Kármán contacted Andrew G. Haley, a Washington, D.C., lawyer who had helped Kármán's sister get a visa earlier that year. Haley was a Falstaffian figure, a rotund businessman known for his ability to get what he wanted, maintaining all the while a mischievous Irish sense of humor. He had developed a great respect for Kármán in their dealings up until now, and the two men shared a natural, unfeigned exuberance. Haley seemed the ideal man to help the rocketeers in their cause. As well as knowing how Washington worked—he had specialized in communications law and had written many of the regulations governing the radio business in the late 1930s—he was also a man touched by romantic enthusiasms. A lover of poetry, songs, and bonhomie, he was more than willing to believe in the dream of rockets. He agreed to help the group form a company that would produce JATOs for the armed forces.

And not a moment too soon. Thrown headlong into war, the navy and air force (formerly the air corps) now jacked up the workload on the rocketeers. Rather more "oomph" was going to be needed than that which propelled the paper-light Ercoupe into the air. They wanted to see 100 solid-fuel JATO units produced, each of which was to deliver 200 pounds of thrust—as opposed to the previous 24 pounds—for 8 seconds. There could be no explosions. If these units could be perfected, then JATOs would prove an invaluable resource for heavily laden fighter planes taking off from aircraft carriers in the Pacific.

The rocketeers were slightly apprehensive about these new demands. As Walter Powell, a new member of the growing rocket team, remembered, up until this point "they'd barely squeaked through without blowing up anything." Indeed, Parsons' fuel—GALCIT 27—was still far too volatile for navy requirements. The success of the Ercoupe tests had in a large part been due to the swift creation and transferal of the JATO units from laboratory to air field. If, however, Parsons could solve the fuel problem, the rocketeers could provide the hundred experimental rockets to the navy while putting their new company in an excellent position to mass-produce and sell more of them.

Hazards, gambles, ventures had become part of everyday life for the group over the years, and they had become as inured to the vagaries of chance as a Greek chorus. Despite the obvious risks, Parsons, Forman, Malina, Karman, and Martin Summerfield each agreed to invest $250 of their own money into starting the company—a considerable sum for students, a teacher, and some rocket tinkerers. All the inventions, techniques, and shortcuts the group had used in their revolutionary work, all the patents they had filed, now became assets of the company, with the founders sharing the stock among themselves.

One of their first decisions was what to call the company. Anything with the word *rocket* in the title was still likely to be dismissed as a joke. When Kármán suggested "Superpower," it was rejected as being too close to "Superman," and, of course, comic book associations of any kind were to be avoided. After much discussion, the Aerojet Engineering Corporation was founded. The purpose of the business was painted in rather broad strokes in the certificate of incorporation: "To design, develop, experiment with, introduce, manufacture, assemble, build, repair, maintain, operate, lease, let, purchase, sell and/or generally deal in and deal with mechanisms, devices and processes now known or which hereafter may be developed, discovered or invented for the propulsion of airplanes and all other kinds of vehicles, projectiles of all kinds and description, or any other machine, apparatus, device or object requiring motive power whether for purposes of commerce, warfare, pleasure or otherwise." No exact products were specified, because as yet there were none. The corporation opened offices on East Colorado Street, in the ground floor room of a former Vita-Juice plant. Helen Parsons planted flowers outside it.

No one trumpeted Aerojet's creation. The headlines of the newspapers the next day spoke of a war department report on the bombing of a Tokyo cruiser, a Nazi spy being sentenced to five years in prison, General MacArthur's promise to wrest back the Philippines, and the plan to transform the Santa Anita racetrack into a camp to hold Japanese and Japanese-Americans who would be forcibly resettled from Los Angeles. Aerojet did not even make the business pages.

The relationship between the new company and the GALCIT project was rather incestuous. Although Parsons and Forman officially went to work full time at Aerojet in order to draw wages, they made frequent visits to the Arroyo, just as Malina and Martin Summerfield regularly visited the Vita-Juice plant. But while the GALCIT Rocket Research Group had now begun to attract Caltech's academic elite—graduates and professors— Aerojet was finding it rather harder to entice suitably qualified workers to help prepare the company for business. Many of the most experienced engineers working for the big aircraft companies laughed outright when they heard of Aerojet's plans, advising the rocketeers not to go fooling with that "Chinese stuff." Unable to entice seasoned engineers to join them, Aerojet did what the Rocket Research Group had done six years before: It turned to the young, bright, and untested. Indeed, Apollo Smith, who had been at the first firings in the Arroyo back in 1936, was lured back to act as chief engineer of the company at the age of twenty-seven. What these engineers lacked in experience they made up for in imagination, and they were more than willing to stretch the rules.

Despite their willingness to disregard convention, many of the young scientists were not quite prepared for the unorthodox working style and perilous "anything-goes" atmosphere that Parsons and the Suicide Squad fostered. With the exception of Kármán and Haley, all the founders were younger than thirty. When they showed up for work, which could be at any hour, they were usually a startling sight. Whether it was Ed Forman arriving, cut and bloodied after his most recent motorbike crash, or Parsons reciting poetry to the secretaries, causing squeals of delight on his every appearance, the company's core scientists seemed not so much unprofessional as insane. One young scientist, Bill Zisch, was so fearful of his future at a company run by such eccentrics that he developed an ulcer from worry. Even Kármán, then president of the company, admitted that during this time Aerojet relied equally on the founders' innate talent for invention and a large amount of verve and momentum— what he called "*Schwung.*"

Such qualms were partially soothed by Andrew Haley. While Parsons worked on improving the JATOs, Haley began traveling coast-to-coast to drum up business for the fledgling start-up. When he was in Washington negotiating with the military agencies in order to fill the company's ever empty coffers, he would send two-hundred-word telegrams back to Pasadena, urging the Aerojet staff in epic tones to "drive night and day with indomitable purpose and inspired leadership." Even his wires, though, were occasionally tinged with the sense of panic that infused the company: "This is the crossroads, with all our reputations at stake."

The shirtless rocketeers back in Pasadena laughed at Haley's rhetoric; they joked that the G in Andrew G. Haley stood for "God." But all shared a sense of urgency during the first formative months of the company. None felt it more keenly than Parsons. His excitement at the great successes of the Ercoupe and A-20A tests had been tempered by the continuing explosions caused by his solid fuels. The solid-fuel JATOs were intended to be Aerojet's first product, and they did not even work properly.

Parsons had established that the GALCIT 27 propellant used in the Ercoupe tests was susceptible to temperature change because of the ammonium nitrate in the mixture. When too cold the nitrate shrank and when hot it swelled, fracturing the solid mass of the propellant charge in the motor. Parsons tried replacing the ammonium nitrate oxidizer with guanidine nitrate, which was not as sensitive to changes in temperature. The mixture, now GALCIT 46, was subjected to twenty-eight days of accelerated temperature changes: For 12 hours the fuel was kept at 100 degrees Fahrenheit, followed by 12 hours at 40 degrees Fahrenheit. Thirty percent of the units still failed. It was an improvement over his more volatile earlier fuels but hardly one which the armed forces would find acceptable. All the variants of conventional black powder seemed to have the same drawbacks. He had to conclude that powder itself was the problem.

Parsons had reached the evolutionary dead end of black powder research. Centuries of tinkering and mixing, of pestle and mortar, of crucible and fire arrow, had left him with a two-foot-long metal tube stuffed with a powder that would explode if left in the cold for too long. It seemed that Goddard and the American Rocket Society had been right—there was no future for black powder rockets. But Parsons insisted on moving forward, perhaps driven by his *Schwung*, or perhaps motivated by the sheer inability to admit that anything was impossible. There was something elemental and primeval about powders, something about their very archaic nature that had always appealed to him, but he knew he had to leave them behind. If the rockets were to fire correctly, something entirely new was needed.

Nothing was to be ignored; nothing was to be ridiculed. Parsons even visited numerous other amateur rocket enthusiasts across the country in the hope of gaining inspiration. On a visit to New York in May 1942, he had watched a particularly poor rocket firing on the banks of the Hudson River. In a letter back to Frank Malina, he was merciless in his assessment:

> His apparatus is crude + he works under considerable disadvantage ... At the start, his valves were leaking, + there was a considerable fume ... Then, as he tried to shut off, the valve stuck, + acid, benzene, fire, etc, raised hell in general. There was no great damage, but that was the end of the test. His measuring apparatus is primitive, and I believe his records are meaningless. He is absolutely foolhardy placing himself + all observers on top of the apparatus, with no protection ... He claims to hold 6 secret patents.

From the comical explosions to the "foolhardy" safety procedures, the letter appears as an almost exact blueprint of the way Parsons himself had worked in the early days of the Suicide Squad. Nevertheless Parsons and his colleagues were now bona fide leaders in their field.

Whatever he thought about the working practices of the novices he visited, Parsons would never entirely transcend the essence of amateurism inherent in his own character. It showed itself not only in his chaotic working practices but in his evident joy in the work he was doing. Despite all the military funding the work had gained, he was still pursuing rocketry with an uncomplicated and eager curiosity. He made up for his lack of theoretical and mathematical schooling with his prodigious memory, broad experimental knowledge, and, most importantly, his ability not to be dissuaded by conventional thought. Black powder would never be a suitable rocket fuel. But the armed forces wanted a solid-fuel JATO. When is a solid fuel not a solid fuel? It was a question that Parsons answered with an Archimedean "Eureka!" As Malina affirmed, it seemed that suddenly, as if by the addition of some ingenious ingredient, "all the various stuff that was in the back of Jack's mind, jelled."

Like all legends the story of his discovery is hazy. Martin Summerfield thought Parsons came by it while driving up to the rocket project; Frank Malina thought it might have come after "communing with his poetic spirits." Others have suggested that it was a collaborative effort between Parsons and Fred Miller, a mechanic, or Mark Mills, a Caltech graduate attached to the project. The most commonly repeated story is that Parsons was walking near the test site when he saw some workmen spreading heavy black asphalt across a roof. The smell and sight of the asphalt seems to have activated his encyclopedic knowledge of myth and historical literature and turned him to thoughts of "Greek Fire."

Greek Fire was a terrifying flaming liquid weapon, the existence of which had been recorded in Sparta as early as 429 B.C. It was most famously used by the Byzantine Empire under the reign of Constantine Pogonatus (A.D. 648–685), whose armies employed the strange burning substance to great effect in their sea battles against the Saracens. The Byzantine sailors would mount bronze tubes onto the prows of their galleys, which emitted jets of liquid fire that set enemy ships alight. The Greek Fire was a potent psychological as well as a practical weapon; it was reputed to be inextinguishable and burned even on water. Because of its devastating effect, the exact nature of its composition was kept secret, passed down from emperor to emperor.

For the ensuing millennia Greek Fire remained one of the great unsolved mysteries of chemistry.

Over the years a number of theories about its ingredients had been advanced: liquid petroleum, quicklime, saltpeter, sulphur, or—most pertinently to Parsons—naturally occurring asphalt. As he stood there, watching the tar being poured, science collided with the romance of ancient history in his unorthodox mind: "Why not get rid of black powder altogether and use asphalt as the fuel?" he suddenly thought.

Parsons rushed back to his laboratory and began work. He melted some common roofing tar to liquid point, then added potassium perchlorate, an oxidizer that allowed the tar to breathe and flame. After mixing, he cast the boiling glutinous substance directly into the rocket chamber, bouncing the casing lightly to fill every possible gap in the motor. Once cool it looked like hardened paving tar, a tough, slightly plastic charge with no cracks in it at all.

The idea of using asphalt instead of black powder overturned centuries of previous thought. Parsons' innovation was strange enough to have come straight from the pages of a science fiction story: A modern-day scientist battling to produce a new fuel is inspired by a mysterious 1,500-year-old combustible to fire a space-age rocket! GALCIT 53, as the mixture became known, performed excellently under tests. It could be stored indefinitely, and it took only five minutes to pour it into a unit, a huge improvement over the laborious forty-five minutes it took to pack a rocket motor with GALCIT 46. Not only that, but it was much safer to handle.

With GALCIT 53 Parsons invented the castable fuel, a whole new category of rocket fuel. It caused a paradigm shift in the world of rocketry, making solid propellants suddenly much safer and immensely more practical. It was, said Malina, one of "the most important discoveries in the long history of solid rockets." The results of Parson's thinking would eventually play a central role in American rocket technology. During the next half-century, GALCIT 53's plasticized descendants would be used to fire the Polaris, Poseidon, and Minuteman intermediate-range ballistic missiles halfway around the world at supersonic speeds. Most poignantly of all, the booster rockets that thrust the space shuttle into space are modeled after Parsons' invention. Although he would not live to see his dream of space travel come true, he was essential to making it a reality.

9

DEGREES OF FREEDOM

Explosion: 1. A violent release of confined energy, usually accompanied by a loud sound and shock waves. 2. The act of emerging violently from limits or restraints. 3. A sudden violent expression, as of emotion.

—*Roget's Thesaurus*

Parsons' confidence was at an all-time high. He had just invented a whole new type of rocket fuel; rocketry itself had been thoroughly validated by the armed forces' continued investment; work at Aerojet was beginning to pick up; and his fortunes were reviving. With the lawyer Andy Haley, he became a roving ambassador for Aerojet, dining with members of the Bureau of Aeronautics and impressing them with the potential of JATOs. He was, in turn, witnessing the rapidly changing perception of rockets within the army and the government. A new rocket project, funded by the navy, had been started in Annapolis, Maryland, and while a visit to it assured him that the GALCIT group had the edge over the newer competitors, he was not averse to picking up pointers from them. "We seem to be ahead in both fields [solid and liquid fuel] in California ... but I have learned much, I have several new angles on storage."

He found time to combine his rocket work with personal errands. Traveling to New York for the first time, he met Karl Germer, Crowley's deputy and debt collector. While in Washington, D.C., he met with Dr. Joseph Auslander, the Library of Congress' poet laureate, and donated some of Crowley's poetical works to the library's book stacks. He was twenty-seven years old and living the high life, eating "turtle soup, sautéed shrimp + crab + ham, half the liquor and all the bluepoints [oysters] in Washington," and, he was pleased to note that because of the war there were four women to every man in the nation's capital. "This is more like life," he wrote to Helen from the East Coast; "activity, excitement, concentration, loneliness, friendship, frustration, fulfillment, drunkenness, hangovers, ups and downs, all blended together in the splendid mosaic of experience." Most excitingly of all, he was taking his first trips by airplane. "I could fly about forever, it's great fun, an exalting experience." He had never seen so much of his country and wrote ecstatically, "Parsons discovers America!"

Parsons' enthusiasm ignored the domestic troubles that had been brewing back home in Pasadena. In June 1941 Helen had gone on a vacation with her mother and one of her other sisters. While she was away, Parsons began an affair with her half sister Betty. She was seventeen years old, ten years his junior. As demure and reserved as Helen had been, Betty was her opposite. Feisty and untamed, proud and self-willed, she stood five foot nine, had a lithe body and blond hair, and was extremely candid—she often claimed to have lost her virginity at the age of ten. An exasperated Jane Wolfe described Betty as having "a quick, deft mind, the spirit to take what she wants, regardless; and I think her chief interest in life is amusement. And she thinks she's so right." Indeed, there was something about Betty that was a little too outspoken, too insensitive to the emotions of normal life. When Helen had confided in her about her stepfather's abuse, Betty had been cold and indifferent; she suggested that this was hardly a subject for Helen to get worked up over. The two sisters had always had a tense relationship. Upon moving into their house, Betty had infuriated Helen by wearing her handmade dresses "as if she owned them." Now it seemed as if Betty was trying to steal Helen's husband from her, too. When Helen returned from her holiday, she found Betty wearing her clothes, claiming to be Parsons' new wife. She was appalled. "Think of it," she recalled years later; "my own sister in my own house with my own husband."

Violent fights erupted between Parsons and Helen. Visitors noticed "cracked doors as well as wooden partitions . . . mute witnesses of [Helen's] combats with Jack." But Helen could not bring herself to leave her husband. Perhaps she remembered all those letters Parsons had written over the years, signed "Semper Fidelis," or maybe she did not want to end a relationship that had survived so many difficulties. Parsons knew he was in the wrong. "Everything is my fault—my own mess," he admitted, but he was unapologetic. "I did it deliberately and would do it again." Indeed, when confronted by Helen directly on the matter, Parsons could be devastatingly blunt. "I prefer Betty sexually," he told her. "This is a fact that I can do nothing about. I am better suited to her temperamentally—we get on well. Your character is superior. You are a greater person. I doubt that she would face what you have with me—or support me as well."

Such merciless candor was encouraged at the OTO, which sanctioned his behavior. For Crowley, marriage was "a detestable institution" and monogamy was "one of the most idiotic and bestial pieces of vanity in human psychology." "People should marry for convenience" he went on, "and agree to go their separate ways without jealousy." From his first visit to the Lodge, Parsons saw that wives and husbands were swapped with impunity. The typical attitude towards such exchanges was nonchalance, such as that expressed by Max Schneider, Crowley's "inside man" at Agape Lodge. In his diary, Schneider dispassionately describes his wife Georgia's post-mass dalliance with another member, Roy Leffingwell: "Georgia calls to Roy to come to bed with her; R. is at first a little surprised, but then follows the invitation without further ado. Good!"

But as much as Parsons was acting in accordance with Crowley's laws, the affair with Betty also fed his own vision of himself. Not only was the affair physically pleasurable and religiously acceptable, to Parsons it was also poetically justifiable. In a letter addressed to Helen from this time, he wrote, "A man, a poet should be alive, be drunken, in love betrayed, hurt, lifted from pinnacle to pinnacle." Looking back on this time a few years later, addressing himself as a mythologized "You," Parsons portrayed his affair with Betty as a crucial step in the inexorable course that would lead to his becoming a great practitioner of magick: "Betty served to affect a transference from Helen at a critical period . . . Your passion for Betty also gave you the magical force needed at the time, and the act of adultery tinged with incest, served as your magical confirmation in the law of Thelema." Nevertheless, despite Parsons' best efforts to live an open relationship with Betty, the plethora of shifting bedfellows could not prevent him from falling in love with her.

However unwilling Helen was to leave Parsons and no matter how stridently Parsons justified his affair through the teachings of the OTO, Helen could not help but feel that he had breached an unwritten contract by sleeping with her half sister. Soon afterwards Helen wrote in her diary of "the sore spot I carried where my heart should be." Naturally more reserved than her husband, she had not embraced the OTO doctrine of adultery as readily. Nevertheless, she now found her own solace in the arms of Wilfred Smith, Parsons' mentor and head of the Agape Lodge.

Parsons seems not to have been bothered by this affair— he had recently written to Helen, telling her to "get Willie a girl- friend"—and if anything the friendship between the two men became stronger. Their personal dramas, though, were caus- ing turmoil amidst the OTO lodge. The older members were frustrated at the "increased laxity" of both Parsons and the new members that he was attracting. They still worried that this young generation was sidelining Thelema and simply indulging their sexual whims. In April 1942, an official meeting was called at Winona Boulevard "because of a dissent from some views of the Order, particularly pertaining to Sex before going up to Mass." Perhaps more than anyone else, Regina Kahl, the blustering opera singer who had been Smith's lover up until this point, was furious at the new state of affairs. Her tantrums reverberated through the house, especially when Smith revealed that Helen was to take over Kahl's role as priestess in the weekly mass. Kahl had helped Smith establish the Agape Lodge in California. Indeed, her drive and domineering forcefulness were all that had kept the lodge from disappearing amidst the expulsions and quarrels of the thirties. But a new wind was blowing; not only had Helen replaced Kahl in Smith's bed, but Parsons was also taking over her role as the dynamic organizer of the lodge. No matter how much members grumbled over sex, when Parsons spoke at group events, many of them were visibly moved by his words. Some of them, with tears in their eyes, likened his appeal to that of President Roosevelt himself.

In New York, Karl Germer, Crowley's second-in-command, was hearing good things about Parsons. Germer had known Crowley in Berlin and had been an early member of the OTO. He had proven a loyal disciple of Crowley, sending him money on a regular basis and suffering greatly for his devotion when in 1935 the Nazis arrested him as a member of an illegal cult and placed him in a concentration camp. He somehow managed to engineer his release in 1940, and he now found himself serving as Crowley's personal representative in America. He was in charge of registering the copyright for Crowley's publications and collecting contributions from members for the Aleister Crowley Publication Fund. Known as Frater Saturnus within the order, his belief in Crowley's work and his devotion to Crowley were absolute. It was widely known that he had married both of his wives not for love but for the monies they could provide to his mentor. In a letter to Jane Wolfe, Germer offered some advice to the new wonder child of the organization. He advised that Parsons "needs a Guru, especially as he lives in Los Angeles' spiritual atmosphere which does not have a good reputation; if he is scientifically inclined, all the better. Let him use most of all common sense and go by his own inner light." He hinted that an introduction to Crowley himself was not far away.

Meanwhile, Parsons believed he had a way to quell the rumblings of discontent in the OTO. He suggested that the group should move from their present lodgings on Winona Boulevard to Pasadena. Not only would the move enable him to devote more time to the OTO, but it would allow them to find larger and more salubrious lodgings if the order was to expand. Smith and Helen backed him up, and the other members assented, especially as Parsons offered a large part of his wages to help finance the move. Through his mother Ruth's contacts, he found a suitable building on the street of his childhood—Orange Grove Avenue.

Prices on Millionaire's Row had fallen dramatically. The old money of Pasadena had been lost in the crash, relocated by the war, or passed on to an altogether different Paradise. The magnificent properties that at one time were valued at $300,000 or more could now be bought for one-tenth of that price. Admittedly the lawns weren't quite as trim as they had once been, nor was the paint on the houses as bright as in the gold-tinted twilight of the 1920s, but the street still exuded the quiet and dignity that signified one was entering an area of rarefied exclusivity.

Nowhere was that character more evident than at 1003 Orange Grove Avenue, the former home of lumber millionaire and Caltech benefactor Arthur Fleming. A man of magnificent philanthropic gestures, he had left his mark on international affairs by buying, refurbishing, and donating to the French government the railway dining car in which the First World War armistice had been signed. In 1940, however, his generosity had been cruelly repaid when Adolf Hitler, in an act of political exorcism, forced the imperiled French government to sign their surrender in the very same railway carriage. After the humiliation was complete, the carriage was destroyed. It was a mortal blow to Fleming. Less than two months after this perversion of his magnanimity, he died, a broken man.

If Fleming had been able to see the new occupants of his house, it might have added insult to mortal injury. On June 9 1942, as Parsons was putting the final touches on GALCIT 53 in the Arroyo Seco, a strange procession wended its way into the old Fleming house. White-haired Jane Wolfe led the way, carrying a picture of Aleister Crowley, followed by Wilfred Smith with *The Book of the Law,* Regina Kahl with the Egyptian tablet, and Helen Parsons with a small model ark. Upon entering the house, Smith, in his most sonorous voice, declared, "Do what thou wilt shall be the whole of the Law."

Jane Wolfe, who had initially been unwilling to move from Winona Boulevard, was suitably impressed by the new building, although she felt somewhat awed by her surroundings. "What heavily restricted Orange Grove will do about us remains to be seen," she reflected. "Perhaps our affrontery in moving in right under their noses will silence them." Parsons seems to have felt no such worries; indeed, the irony of renting 1003 must have amused him intensely. It was in this very house that Fleming had entertained the greatest minds of Caltech, including Albert Einstein himself. Now Parsons, so long kept on the margins of Caltech's social and scientific life, was claiming the home of the institute's greatest benefactor as his own. Perhaps for him the move was also a return to the personal playground of his youth, to the environs which had spawned his world of fantasy. He, Helen, and Betty moved in immediately; indeed, such was Parsons' rush that he left many of his chemicals and explosives behind at his old house on Terrace Drive, much to the landlord's alarm.

The house on Orange Grove had stood empty since Fleming's death two years earlier, but many of its splendors remained
intact. Built in 1899 for Fleming and his wife by the architect
Frederick Roehrig, it was one of the first Craftsman-style houses
to appear in Pasadena. Upon entering the house, the visitor came
first into a large hall, boarded with ornate cedar panels and dominated by a sweeping, hand-carved, mahogany staircase, over
which Wolfe had hung the portrait of Crowley. The original silk
curtains still hung in the windows on either side of the front door.
The hall opened into a large oval dining room, with woven tap-
estry walls above birch paneling; the adjacent music room housed
Regina Kahl's piano and Parsons' phonograph and records. A
giant, carved fireplace held court in the long living room, from
behind which led a secret passage, one of many in the house.
Glass doors led outside onto a large terrace. Here, it was decided, lodge meetings would be held and the mass performed.
But the grandest room of all was the library, its four walls lined
with bookshelves. Above them the walls were covered with
tooled leather depicting scenes from German village life. Admir-
ing these scenes, the visitor might almost fail to notice the gold
leaf flaking off the ceiling. The house's large kitchen was to become the center of daily life in the new Agape Lodge. It opened
onto a covered porch off the building's back, where day meals
were taken. Stairs led down to the cellar, a space large enough
"to hold 50 people easily" but which was to be devoted to holding the OTO's sizeable collection of wine.

On the second floor were six bedrooms, each with a tiled
bathroom. Rooms were allotted according to the OTO's hierarchy. Parsons and Betty shared a two-room suite, an arrangement
that afforded them a modicum of privacy if they chose to have
separate nocturnal affairs. Parsons covered the walls with knives
and swords which he had collected over the years. "He was not
happy unless there was a knife in the room," remembered one
of the other lodgers. While these signaled his machismo, other
ornaments bespoke of his mystical nature, such as the statue of

the Greek god Pan standing in the corner of the room. The master bedroom was reserved for Wilfred Smith and Helen. On the top floor a further five rooms were found, former servant's quarters, reserved for the less exulted and lower-rent-paying members of the Lodge.

Slightly apart from the main house stood the garage and laundry, a two-floor structure, in which Parsons chose to build himself a new home laboratory. Twenty-five acres of garden interlaced with tiled walls stretched down towards the Arroyo. Walking along one of the paths, one came upon a pergola with mosaic tile floors, and a glass tile fountain ten feet in diameter, which was used by the lodge's children as a paddling pool. A large structure known as the Tea House kept watch over the bay trees at the bottom of the garden; it would be a favored location for the assignations, both secret and known, of the household. Parsons would often take the children of the house to search for rings in the grass, which he said were left by dancing fairies (in fact, caused by the unchecked growth of the fairy ring mushroom which grows in circles), convincing the children that the garden was enchanted. And if this bounty proved insufficient, just over the walls lay the charmed rambles of the Busch Gardens.

It was a house fit for a king, but it cost only $100 a month. Parsons, who had grown up surrounded by conspicuous wealth, now attempted to replicate that life. At 1003 Orange Grove he promoted both epicurean and intellectual enjoyments. He would fence with his friends on the grounds before retiring to partake of his treasured collection of fruit brandies in the kitchen. In this mansion he could both womanize and practice his magick with ease. He would create a lifestyle that echoed the Byronic heroes of the past, resounding to the ever present texture of myth, while chiming with the teachings of Crowley's new aeon.

Along with Kahl and Wolfe, the other OTO members who moved into 1003 were a diverse group. Fred and Grace Miller were both old hands in the burlesque trade, but while Grace was still a dancer in various traveling companies, Fred had taken to working in a shipyard to pay for the upkeep of their two young sons, who lived in the house and more often than not were left to be looked after by Betty. Frederic Mellinger was a former avant-garde actor and writer from Berlin. Other German expatriates had helped him find occasional bit parts in Hollywood movies, but most of the time he was unemployed. Believing housework was below him, he confined himself to drawing up astrological charts for the members of the house. A tall Lapp called Jonas Erickson, an old friend of Wilfred Smith, also lived at the house. An epileptic, he was prone to having fits, terrifying the children of the household. Deeply loyal to Smith he acted as the house's strongman, chopping lumber and carrying it to the house. Phyllis Seckler, wife of the jailed Paul Seckler, also stayed here, on the top floor with her two children.

The house and gardens badly needed care. A bees' nest was found on the porch and an infestation of bats discovered in the house's attic. The water pipes had a habit of bursting. Jane Wolfe named some of the necessary tasks in her diary: "screws, electric connections, sawing, adjusting, watering, clipping hedges, planting the rose bushes we brought with us, scouring, scraping." The members fenced off a large vegetable garden and planted beets, carrots, beans, and tomatoes. They bought rabbits, chickens, and goats to keep for food or to sell at market, or, if need arose, to provide animal blood for rituals and for the Gnostic Mass's cakes of light. Parsons' mother, Ruth, occasionally came to visit and to help him with the household chores. Unfortunately, what she thought of her son's strange friends and surroundings is not known.

The group imagined that they would share all responsibility; they envisioned 1003 as a "Profess House," a utopian mission sheltering a community of true Thelemites. Inevitably, the reality of commune living came as a shock to all. The elderly Jane Wolfe was left to do the shopping for the entire house, while Phyllis Seckler was expected to do all the cooking as well as look after her own children. Mellinger's duty was to water the garden, but he often slunk off from work and it began to die. Smith—who had given up his job as a gas clerk—was asked to look after the animals and the vegetables, jobs that he complained were more in keeping with those of a handyman than of a priest of the OTO. Meanwhile, many felt that Betty was shirking her share of the labor. The incessant cries of the members' children added to the tension; the noise sent the usually benign Parsons into the foulest of moods. Throughout his years of marriage with Helen, Parsons had insisted that he never wanted a child and had practiced coitus interruptus to prevent this.

Despite all of these difficulties, Parsons was dedicated to making the move work. He made an agreement with Smith and Helen (who was acting as the organization's treasurer) that he would hand over his entire salary to the Church of Thelema to help with the running of the house, keeping just $10 a month for himself. Even though Parsons was the only member of the house to make such a financial commitment to the Lodge, he never complained that his generosity was being abused. He would provide money for the OTO, and the OTO would provide a family for him.

By the end of 1942 that family was growing. Over forty new members had been initiated into Agape Lodge, largely thanks to Parsons, who was proving remarkably successful in attracting people to join the order. His unrestrained enthusiasm for Crowley's philosophy was evident, as a letter to a holidaying Ed Forman showed:

> I'm a great one to talk, but it's about time you got out from under the domination of your emotions . . . you should be able to say to your temper, your hang-over prejudices, your old conflicts and impulses, "I am master here—you do what *I* say"—and make it stick. It would be fine for all of us to learn more tolerance towards others, and more intolerance towards ourselves . . . above all things have a good time. You only live once.

It was not long before Parsons had enticed Forman and his second wife Phyllis into becoming members, along with his old friend Tom Rose from the Hercules Powder Company, and Barbara and Richard Canright, mathematics experts from Caltech who had worked on some of the JATO tests. "Jack does more than any one trying to interest outsiders in the field," wrote Jane Wolfe to Germer. With much of California swamped with soldiers and sailors on recreational leave, the region was infected by their spirit of carpe diem. Parsons thus found a receptive audience for his recruiting drives. The Agape Lodge promised an escape from the world's worries. "The only reason people would have joined was out of fun—because it was kind of 'far out'," remembered Forman's third wife, Jeanne. "It was a time in a society when everything was new & different—and out there and strange and marvelous!" Taking their cue from the strange interfamilial relations of Parsons, Betty, Smith, and Helen, many of the new members and residents of 1003 started affairs of their own. Fred Miller began sleeping with Phyllis Seckler, as did Smith on occasion. Miller, thirty-eight years old, also began an affair with Jane Wolfe, by now sixty-seven, and the two began performing sex magick rituals together. Despite the war work Parsons was involved in, despite the reports of the Pacific fleet repelling the Japanese at Midway and of Allied convoys fighting their passage across the precarious Atlantic, the war seemed a very long way away from 1003.

The Thelemites planned an equinox party for the autumn of 1942. They invited friends and colleagues from Caltech and Aerojet to mingle with OTO members in the lavish setting of their new home. It was to be a secular party, with no preaching, but in the days prior to it Parsons became somewhat anxious about bringing these two parts of his life together. Bypassing Wilfred Smith, Crowley had written to him that he should take over some of Smith's duties, notably the distribution of Crowley's new manifesto for the OTO called *Liber Oz*. Crowley believed that this work would finally boost Agape Lodge's membership in California, something he no longer thought Smith could achieve by himself. It was a great honor to receive a personal letter from Crowley, but Parsons was disturbed by the manifesto's contents.

Little more than a pamphlet, *Liber Oz* professed the familiar Crowley doctrine: "Man has the right to live by his own law, to live in the way that he wills to do: to work as he will: to play as he will: to rest as he will: to die when and how he will." But unusually, it went further: "Man has the right to kill those who would thwart these rights." If one of Aerojet's military associates should see a copy of this at the party, Parsons would likely have been judged unfit for his job. He asked the other members to remove all evidence of *Liber Oz*. His request was something of a disappointment to those who had considered him the order's great new proselytizer, but Parsons recast his caution as evidence of his grand vision. He told them that he was not being censorious, nor did he disapprove of Crowley's work; rather when he did become a martyr or get sent to jail, he wanted "to make headlines, and at present such an action would not cause a ripple in the public consciousness." The gesture was typically romantic and self-aggrandizing, but Parsons succeeded in winning the long-standing loyalty of many members of the house, including Jane Wolfe. "He knows his power, he knows what he wants, and he knows his royalty," she wrote. The party was a success, ending in a drunken dance at the pergola. One by one, OTO member or not, the guests stripped off their clothes and cavorted around the fountain naked.

Parsons' caution over *Liber Oz* was not groundless. Soon after the party, the Pasadena police received a letter from San Antonio, Texas. Signed "A Real Soldier," it stated that a "black magic" cult flourished at 1003 Orange Grove Avenue, practicing "Crowleyism" and "Sex Perversion." The letter also named one of the house's inhabitants, Frederick Mellinger, as an "enemy alien." When FBI officers came to investigate, Parsons gave them a tour and smoothly explained that 1003 housed a fraternal study group who discussed "philosophy, religion, personal free-dom and the mysteries of life and eternity." In the ensuing years the police and the FBI would investigate 1003 time and again, but they would never find anything incriminating. Nevertheless the FBI did open a file on Parsons, detailing his link to a "love cult." By the end of his life, the file would stretch to nearly two hundred pages.

Within the house angry members pointed fingers of suspi-cion as they debated the identity of this "Real Soldier." Many assumed that the still-jealous Regina Kahl or one of her scan-dalized relatives had sent the letter. Kahl had gone to Texas to recuperate from a persistent illness, which had been exacerbated by the lodge politics and the hard work of communal living. However, any number of disgruntled souls might have informed on the Lodge. Parsons' science fiction friend Grady McMurtry, for one, had become upset by events at 1003. Soon after Mc-Murtry and his wife Claire had been initiated into the OTO, their marriage had disintegrated as Claire gave herself over to the licentiousness of the Lodge, sleeping with both Parsons and Smith. Parsons eventually helped to pay for an abortion when Claire discovered that she was pregnant (something he would also do later for Betty.) McMurtry was shocked by his former idol's callousness.

Crowley tried to patch up the two men's relationship, ex-horting McMurtry to fight against the jealousy he felt. Even so, with his marriage destroyed and now himself drafted into the army, McMurtry could make only infrequent trips back to 1003. When he did, his visits filled him with unease. He had been rather surprised, when talking to Smith, to find that the mass had not been performed since the OTO had moved into 1003. From his army base he wrote Parsons a letter describing the group which now hung out at 1003 as "a bunch of empty-headed Athenians" (a phrase he pointedly chose from Crowley's *Konx Om Pax*) and he described Parsons as "being coked up like a snowbird by Wilfred."

Although McMurtry seems to have meant that Smith's influence on Parsons was narcotic, his former friend was indeed indulging in various drugs by this time. A regular drinker of home-brewed absinthe and an occasional user of marijuana, he now appeared to have turned to cocaine and amphetamines to cope with the pressure of work at Aerojet and GALCIT. His colleagues noticed he was soaked in a perpetual sweat during the day, a common side effect of amphetamine abuse. "Jack was an expert on drugs," recalled Robert Cornog, a friend from the time; "he knew them backwards and forwards." It is possible that he had begun to experiment with making his own drugs in his home laboratory. Such substances were not intended purely for recreation. Crowley preached that drugs could aid in the enactment of magical rituals and facilitate "astral travels" when the magician supposedly leaves his body and communes with the universe and the beings who inhabit this higher plane.

When news of the party, the FBI investigation, the drugs, and the constant partner swapping got back to Karl Germer, he saw the turmoil as proof of Smith's bad influence. Breathless with rage he wrote to Crowley:

> I'd better slow down. I'm getting hot under the collar. Can't even type straight. However, in American slang, "it burns me up" to see the Order, and all it stands for, represented as a Seventh Ave Social Club, with "who slept with whom" vieing with drunk exhibitionism as the topic of the moment ... I am no prude, but neither do I care for obscenity and vulgarity in my sexual life. Nor do I think it a good foundation on which to build a stable, worthwhile Lodge of the Order. Do you?

It seemed a strange question to ask of Crowley, a man who had in the past purposefully rejected every moral standard that Germer now invoked. Of more relevance was the fact that these debaucheries seemed to come at the expense, literally, of the

Order—Crowley's monthly contributions were much lower than he wanted them to be, especially as he had embarked on a new and vital project. With the help of one of his British disciples, Crowley was designing a deck of tarot cards rich with astrological, numerological, and cabalistic symbolism. "This will I think rank as the most important thing I have done in my life," he wrote. He looked to America and the Agape Lodge for the financial support to realize this project, but with Smith at the helm of Agape Lodge, such aid was not materializing.

In January 1943, Crowley ordered Smith to step down as head of Agape Lodge until Smith had undertaken "some definite personal action, conceived and executed by [himself] alone, to the advancement of the Work of the Order." Parsons was confused by this demand. Smith had initiated him not only into a religion but also into a very enjoyable sexual lifestyle. What's more, Smith's charm and sagacity was keeping relations cordial between him and Helen. Parsons and Helen wrote a joint telegram to Crowley, insisting that he had been misinformed about Smith: "Remember we have hard job cannot long afford to lose Wilfred. Things have developed well . . . Do not believe everything you hear. Love and trust. Jack and Helen." They needn't have bothered. Smith, the focus for all Crowley's disappointments in the failure of his order, was a lost cause.

Agape Lodge was divided between those who backed Smith and those who wanted change. Parsons vacillated. He had always been slightly naïve and susceptible to forcefully made arguments. Whoever held his ear last had his support, at least until his enthusiasm wore off. The standoff between Smith and Crowley recalled the situation Parsons had found himself in ten years before at the University School, when he had been bullied into supporting his recalcitrant schoolteacher against the school's headmaster. Would he side with the rebel or the master?

Hoping to restore Smith to Crowley's favor, Parsons suggested that he and Smith found a weekly magazine. They would call it the *Oriflamme*. Filled with articles written largely by

Parsons and Smith, it would show their devotion to the Great Work of spreading Crowley's gospel. Despite these good intentions, in Parsons' hands the *Oriflamme* expressed not Crowley's enthusiasms but his own. The first issue's centerpiece was an untitled poem written by Parsons, a rip-roaring hymn to debauchery that seemed to reveal the true reason for the order's existence:

I height Don Quixote, I live on Peyote
        marihuana, morphine and cocaine.
I never knew sadness but only a madness
        that burns at the heart and the brain,
I see each charwoman ecstatic, inhuman,
        angelic, demonic, divine,
Each wagon a dragon, each beer mug a flagon
        that brims with ambrosial wine....
The mountains are palaces, women are chalices
        meant to be supped and not sold,
The desert a banquet hall set for a festival,
        ripe for the free and the bold;
The wind and the sky are ours, heaven and all its stars,
        waken, and do what you will;
Break with this demon spawn'd hell-inspired nightmare
        bond—Magick lies over the hill.
They said I was crazy, ambiguous, lazy,
        disgusting, fantastic, obscene;
So I hied for my sagebrush and cactus and corn mush,
        To see if the air was still clean.
Oh, I height Don Quixote, I live on peyote,
        marihuana, morphine and cocaine,
And may I be twice damned for a bank-clerk or store hand
        if I visit the city again.

If Parsons thought that this poem would improve Crowley's view of Agape Lodge's operations, he was quite wrong. Upon receiving his copy, Crowley wrote angrily to Jane Wolfe, "What could have been better calculated to revive the ancient stories about drug-traffic and so on!"

Crowley could be devastatingly contrary at times; after all, his *Liber Oz* had recently urged the slaughter of those who would deny man his Thelemic rights. But to his followers he was as infallible as the Pope, and they had to accept his many contradictions without undue fuss. It was true that he wanted his disciples to be enflamed with passion, but this passion was to be focused on Crowley himself. "Jack's trouble is his weakness," wrote Crowley; "and his romantic side—the poet—is at present a hindrance." Crowley doubtless also resented the fact that while he was living in a small apartment in blitz-hit London, far away in the safety of California the residents of the luxurious 1003 Orange Grove enjoyed a Dionysian climate of excess.

The embarrassed and frustrated Parsons quickly shelved the *Oriflamme*. It had done nothing to reduce the hostilities at 1003. No one dared to argue with Crowley himself, so they blamed Karl Germer and Max Schneider for fueling his hostility towards Smith. All agreed, however, that Parsons was the only alternative. Though he was young and Smith's protégé, he was intelligent and most of all a good advertisement for the OTO. And even he had begun to feel a little restless under Smith's leadership. He was by now champing at the bit to perform magick that could produce distinct psychic phenomena. But Smith urged him to slow down, to keep a "magical diary" documenting the more basic rituals he undertook, to hold to a steady progression through the degree system of the OTO. The impatient Parsons, though, had already started performing magical rituals far in advance of his actual grade in the order. As in his rocket work, he left theory behind in the wake of his will to experiment.

By early 1943 the internal feuding of the past months, not to mention the strain of the free sexual relations, had caused an exodus of 1003's original residents. Only five Thelemites now lived in the house—Parsons, Betty, Helen, Smith, and Jane Wolfe. Parsons began to invite friends from Pasadena and Caltech to

fill the vacancies. Although Smith continued to protest his innocence, Parsons was granted temporary control of the lodge by
Crowley, and Germer gave him some secret magick documents
to ensure his allegiance. Although Parsons still supported Smith,
he was growing increasingly interested in creating a direct relationship with Crowley, whom he sent $300 of his own money
towards the publication of the tarot. Soon afterwards, he extended an olive branch of reconciliation to Max Schneider,
Smith's greatest critic. Schneider was pleasantly surprised, and
after sitting in on a lodge meeting, he reflected that "Jack is
doing very well as Master, he has poise and dignity."

With Smith relegated to the sidelines, Parsons began organ-
izing frantically. He started a tarot study class, a Crowley study
class, even a bridge class, in order to attract new faces to the
fold. Magical attainment would be all important, though the
bacchanalian lodge parties would continue. The minutes for
the weekly lodge meetings, meticulously kept by scrupulous Jane
Wolfe, offer a snapshot of the OTO's activities at this time:

> Sister Forman [Ed Forman's second wife, Phyllis] asked
> about the Second Yearly Party. Saladin [Parsons, who had
> assumed Smith's previous Lodge name] designated June
> 22nd for a Solstice Party, saying it should be small & se-
> lect, only possible candidates invited ... Brother Canright
> proposed stressing the fraternal side of the Order, benefits,
> sports, a Club House, rather than the philosophical side.
> Saladin suggested Badminton, Fencing. Also that the Sec-
> ond Degree should be pushed through that there might be
> a special group who would push on to the Third Degree.

The 1943 summer solstice party attracted more than forty
people, and certainly gave a taste of the benefits of lodge life.
Parsons gave a short talk on the OTO, but the dancing and re-
freshments really got things underway. At around midnight a
candlelight procession led by Parsons and accompanied by beat-

ing tom-toms wound through the grounds and finally stopped at the pergola. Here couples leapt over a fire hand in hand. Fire dancing was a traditional pagan fertility ritual, and during 1943 five female members of the Agape Lodge had children. Helen was one of them. She gave birth to a boy, Kwen, in April of that year. Although the child bore Parsons' name, no one was in any doubt that Smith was the father.

Parsons had asked Crowley not to take any further action against Smith until Helen had given birth to his child. Once Crowley heard news of the birth, however, he initiated an ingenious scheme to rid Agape Lodge of the troublesome Smith forever.

Not long after Kwen was born, Crowley sent Smith a twelve-page letter entitled "Liber 132," after Smith's magical number. It was a work of byzantine complexity, presenting the most abstruse mixture of cabbalistic calculation, omen telling, and oracle reading Crowley could muster, and including Smith's horoscope, drawn up and cross-referenced with some broad readings from *The Book of the Law*. At its end Crowley declared that his computations could only point to one thing: "*WIL-FRED T. SMITH is not a man at all: he is the Incarnation of some God.*"

With a fanged smile of satisfaction peeking out from behind the solemn deadpan intonations, Crowley declared that Smith's mission was to find out which god he was the incarnation of. In order to do so, he would have to leave 1003, have his forehead tattooed with the number 666, and journey into the desert to contemplate his task. Most importantly, however, he was to have no contact with other members of the Lodge because "the divine nature must never be contaminated or cheapened by human associations."

The members of Agape Lodge were struck dumb. No one could tell if the letter was a blessing or a curse. The calculations were precise, but one had to wonder whether the systems had been somewhat abused. Nevertheless, Crowley's word was gospel. As for Crowley, he would insist that "Liber 132" was a most serious and genuine scripture—at least when it suited his purposes. In his more generous moments, he even claimed to believe that Smith could be redeemed. Writing to Max Schneider shortly after the event, however, he was quite open about his plan. "No doubt by this time you will have got my solution of his [Smith's] problem," he commented. "His departure should clear things up considerably, although it will take a little time to get rid of the old influence." Smith hadn't been deified; he had been framed. Writing to Crowley, the defeated Smith seemed to accept his fate. "Many times these many years I have speculated as to how and when my turn would come as it has so many others, and now it is here."

Smith did not go quickly. He had an infant son and Helen to look after. To follow through with Crowley's orders would have meant literally becoming a hermit and being exiled from the comforts of 1003 forever. But the more he procrastinated the more tensions in the house rose. Betty and Helen were feuding more than ever, with Betty telling her half sister, "Why in hell don't you get out of here? This is no place for indigent mothers and bastard children." Parsons was himself confused and angry at this dual pull on his emotions. After innumerable false starts, Smith, Helen, and their son eventually left and traveled to OTO member Roy Leffingwell's turkey farm in the desert to begin his magical retirement.

Crowley's cunning plan almost left him without a new leader. Parsons, feeling cut adrift from Smith and fatigued by his unrelenting rocketry work, tendered his resignation to Crowley almost immediately after Smith had offered his. He had had enough of the infighting, of Crowley's "appalling egotism, bad taste, bad judgement, and pedanticism," and he declared himself content to go it alone for a while. It took all of Crowley's serpentine skill, paternal charm, and out-and-out bullying to tempt him back. Telling "Dear Jack" that he was writing to him "as an elder brother and true friend," he refused to accept

Parsons' resignation. "I hope that you will think everything over very carefully and make up your mind to continue to bear the great responsibility and deserve the great honour which is yours," he wrote, playing up to the self-aggrandizing element in Parsons' character.

Parsons was not blind to Crowley's tricks, but he was still besotted with the man's philosophy and magick. He withdrew his resignation. For a long time, Crowley had been a remote figure, his presence diluted and fractured through the anti-Smith scheming of Schneider and Germer. Now Parsons gained a much closer relationship with the man whom he would begin to call "father." For the twenty-eight-year old Parsons, Crowley was the latest in a long line of father figures, not to mention the greatest exponent of magic of the age, while Crowley needed the Pasadena lodge to keep funding him and saw in Parsons both a wealthy benefactor and an enthusiastic novice. Maybe this disciple, young and intelligent, would succeed where so many others, most recently Smith, had failed.

While Parsons spent his evenings trying to restore order to the OTO, he was spending his days at Aerojet's new premises with Forman and Summerfield, assisting in the transition from experimental to full-scale production of solid- and liquid-propellant rocket engines. This was not easy, for their equipment was basic at best. Creating Parsons' asphalt-based fuel was a labor-intensive job; the asphalt had to be sliced with an axe before it was added to the melting pots, where it required constant stirring by hand with large wooden paddles.

Aerojet was very much a hand-to-mouth concern. The company had plenty of orders but was struggling to fill them. Profits from one order paid for work on the next order. New contracts were desperately sought for the embryonic business, and all the members of the old Suicide Squad now found themselves, like Parsons before them, as unlikely ambassadors, traveling the country to persuade aeronautic companies and the military of the usefulness of their product. They experienced a few upsets along the way. Parsons arranged a demonstration of their JATOs aboard the aircraft carrier USS *Charger* in Norfolk, Virginia. Representatives from the navy took their seats along the side of the deck as a Grumman fighter plane fitted with solid-fuel JATOs was prepared for takeoff. When Parsons gave the order to fire the rockets, the Grumman left the deck with a roar. Then, in Kármán's words, "Something happened that we had not considered. The rockets discharged a billow of smoke right on the assembled Navy officers and turned their neat blue uniforms into a dusty yellow. One man rose and blustered that the device was impossible. Most of them, however, took it good-naturedly, but said that we should see them again when we had found a way of getting rid of the smoke."

Aerojet worked at dizzying speed on both research and production. "Everything is going so fast around me that it is bewildering," wrote Malina. "The company will either be a wonderful success or a glorious flop, there won't be any in between. Practically all the old gang is working from 10 to 12 hours daily." When the army air corps asked for 2,000 JATOs by the end of 1943, all hands were called to help. Aerojet by now had some one hundred employees, but even Andrew "God" Haley could be found in his overalls, working late into the night, rolling JATOs across the floor so that the order could be finished in time. The company managed to complete it in the early hours of the New Year.

Because of Aerojet's proximity to Caltech, Kármán's affiliation, and its undeniable success, many of the professors and mechanics at the Institute were now keen to be taken on as consultants. Fritz Zwicky, who six years before had dismissed the Rocket Research Group out of hand, was brought in as head of the company's Research Department. Even Clark Millikan pleaded with Kármán to let him in on the company. The GALCIT operation was supposed to concern itself with the research and development of rockets, and Aerojet was meant to concentrate on the manufacture and sale of JATOs. In actuality it was hard to differentiate between the two operations; the work overlapped so much that Parsons, Forman, Malina, and Summerfield moved freely between them.

Like Aerojet, the GALCIT project kept expanding, even if one still had to drive across the dry bed of the Arroyo to find it. The style of the buildings at the Arroyo had not changed, but each week it seemed as if another building popped up from the dirt. For visitors, walking past these rickety shacks, knowing that a particularly hazardous experiment was going on, was one of the most frightening experiences of their lives. The air force wanted to develop missiles that could be launched from a plane and propelled at high speed underwater—hydrobombs—so a towing channel 500 feet long, 12 feet wide, and 16 feet deep was also under construction. The rockets tested in the channel would be the largest yet fired by Parsons' asphalt fuel. Malina wrote back home of the changes. "There is one new aspect to the situation—whereas not long ago a handful of us worried about *all* the problems, now each problem has a handful to worry about *it*. That, I suppose, represents some progress."

Malina ran operations at GALCIT in the Arroyo. He shared an eight-foot-by-ten-foot office with his secretary, Dorothy Lewis, and his technical aide, Gene Pierce. "We had two desks, I had a chair," remembered Pierce. "I could use the desk when Frank was out. If we had a visitor I would get up and go out." Working conditions were so impractical that most decisions were made, as Martin Summerfield recalled, underneath a nearby tree. The test pits were adjacent to the buildings, so when a rocket was fired, as happened every few minutes, all conversations had to stop. When the roar subsided, it was replaced by the intense buzzing of the building's switchboard, as residents along the Arroyo furiously phoned in their latest noise complaint. For a few minutes after the firing many of the secretaries refused to go outside; the toxic fumes from the liquid-propellant rockets had a nasty habit of melting their nylon tights.

Despite the high security that now surrounded it, the GALCIT project still preserved the familiar chaotic mood of the old Suicide Squad. Dorothy Lewis remembered a hardworking but relaxed family atmosphere. "Jack Parsons was the clown of the group. He loved to play jokes." As in Parsons' youth, these jokes consisted mostly of strategically placed explosives. A safety engineer, a large retired army man, was one of the more obnoxious new recruits to the GALCIT project, forever questioning Parsons and the others about the safety of their operations. The rocketeers decided that he should really be given something to worry about. They hid smoke bombs in particularly hazardous testing pits, and then, when everybody was eating lunch, detonated them one after the other. The shocked security engineer rushed frantically out of the canteen into a vast cloud of smoke from what seemed like an array of exploding test cells. Of course, the pranks, like the rockets, would sometimes backfire. When firecrackers were placed in a tin near some test pits holding highly explosive liquid propellant, the blast didn't just fool the reviled safety engineer. Many of the mechanics working on the engine feared that months of research had been destroyed. They weren't pleased to find out it had just been a joke. And when he wasn't sabotaging the test pits on the GALCIT project, Parsons could be found peering into the offices, looking for pretty young secretaries who were doing their war duty at Caltech and who might be enticed back to the delights of 1003.

10

. . . .

A NEW DAWN

The hours darken and the years
Grow black with evil things
And mad machines spawn monstrous fears
That follow sleep with sombre wings.

—JOHN WHITESIDE PARSONS

Outside the town of Los Alamos, New Mexico, the desert was once again being stretched out for use as a scientific canvas. Since 1942 a burgeoning construction facility had been stationed here, its aim the creation of a bomb of unprecedented power. Under the charismatic leadership of J. Robert Oppenheimer, an extraordinary group of American and European refugee scientists had been gathered to work on what was known as the Manhattan Project. An impenetrable cloak of secrecy surrounded the project, or so the Los Alamos authorities thought. They were disconcerted when they were handed the March 1944 issue of *Astounding Science Fiction* magazine.

Between stories of parallel worlds and galactic battles appeared a story called "Deadline," written by Cleve Cartmill, an LASFS member and occasional visitor to OTO ceremonies. Set on an earthlike planet, the story described the adventures of a commando, albeit one with a prehensile tail, assigned to destroy a uranium-fueled bomb held by a Nazilike power before it could be used in war. The description of the fuel was curiously prophetic. "U-235 has been separated in quantity sufficient for preliminary atomic-power research and the like. They get it out of uranium ores by new atomic isotope separation methods; they now have quantities measured in pounds ... They could end the war overnight with controlled U-235 bombs."

Cartmill's story was packed full of technical data, which the authorities presumed could only have come from a leak at the Manhattan Project itself. It was not long before military intelligence came to question Cartmill and his editor, John W. Campbell. Campbell—the story had been his idea—assured them that all the details were readily available in the public domain and that he had woven them together with his own scientific imagination; like most science fiction editors and writers of the time, he had a degree in science—from MIT no less—and he was well aware of Otto Hahn's and Fritz Strassman's discovery of nuclear fission in 1938. While the eventual revelation of the atomic bomb in 1945 would astound the world, in Campbell's eyes it would vindicate science fiction's role as prophetic literature.

As the worlds of science and science fiction continued to coalesce, so Parsons himself was spending more time in the company of professional science fiction writers. He had become a regular guest of the Mañana Literary Society, a group of authors who met at the Laurel Canyon home of the writer Robert Heinlein. Lean and intelligent, with a pencil-thin moustache and a penchant for ascots, Heinlein was swiftly winning fame as the preeminent writer in science fiction. When Parsons first met him in 1942, he was thirty-five years old, having begun his career late as a science fiction writer. He had spent some time in the navy before being invalided out of the service, and he had gone on to a brief career in politics, which culminated in a run for the California State Assembly as a member of Upton Sinclair's left-wing EPIC (End Poverty in California) movement. Beaten by the Republican incumbent, he turned to the pulps for quick money and now, along with Isaac Asimov and L. Sprague de Camp, was rewriting the rules of the traditional science fiction story, crafting realistic characters and beautifully wrought depictions of the future. His stories of lone geniuses as proficient with their fists as with their slide rules proved remarkably popular in the pages of the pulps, and his fame would only increase in the postwar years when he published such best-selling novels as *Starship Troopers* and *Stranger in a Strange Land*.

Heinlein met Parsons at a meeting of the Los Angeles Science Fiction Society. As one of the earliest members of the American Interplanetary Society, the first rocket society in America, he was impressed by the freethinking rocket scientist and invited him to the Mañana society, so named "for all the stories that would be written tomorrow." Here some of the finest science fiction pulp writers of the time met to drink cheap sherry and talk over new stories.

Those present included William White, also known by the name Anthony Boucher and a hundred other pseudonyms, whose murder mystery stories were infused with Catholic iconography. Cleve Cartmill, a beat reporter crippled from polio, whose atomic bomb story would so alarm the Manhattan Project, had already met Parsons and visited the OTO on Winona Boulevard. Jack Williamson, who also knew Parsons, could be found in a corner of the room, mulling over stories that mixed parallel time streams with Amazonian dominatrices from the future. Visiting guests from outside Los Angeles would also appear at the society's meetings: L. Ron Hubbard, who could type 2,000 words an hour without revisions and who seemed like a character in one of his dizzying tales of psychic powers and strong-jawed supermen; the film director Fritz Lang, whose early German pictures *Die Frau im Monde* and *Metropolis* had been two of the earliest science fiction films; and Lang's fellow German ex-patriot Willy Ley, who was now doing his best to popularize the idea of space travel through his factual science articles in the pulps *Astounding Science Fiction* and *Amazing*

Stories. Into this eclectic group came Parsons, with his talk of rocketry and magick.

Heinlein's interest in Parsons might have influenced two of his stories from this period, "Magic Inc." (originally published as the Crowleyesque "The Devil Makes the Law") and "Waldo." Both drew heavily on *The Golden Bough,* and both spoke of a future in which magic had become just another branch of the sciences, a common tool like electricity. Parsons' influence at the Mañana meetings is even more evident in Anthony Boucher's novel *Rocket to the Morgue,* which took the society as its subject.

Rather like Parsons' and Malina's own prewar roman à clef, in which the Suicide Squad appeared as fictional rocket scientists, the members of the Mañana society appear in Boucher's book as science fiction writers. Parsons (or rather a composite of Parsons and Edward Pendray, the head of the American Rocket Society) becomes, in the novel, Hugo Chantrelle, an "eccentric" Caltech scientist whose interests include "those peripheral aspects of science which the scientific purist damns as mumbo-jumbo, those new alchemies and astrologies out of which the race may in time construct unsurmised wonders of chemistry and astronomy." Boucher's portrait of Chantrelle suggests that the symbiotic relationship Parsons enjoyed with the science fiction writers was more gratifying for him than his relationship to Caltech. He writes, "They [the science fiction writers] received his heterodox views on the borderlands of science far more courteously than did his laboratory associates." In the novel, the thinly disguised characters lurch through a plot in which bombs are mailed, people are stabbed, and a crime-solving nun flutters around the crime scenes. Chantrelle/Parsons' rocket appears at the book's end as the final and most gruesome murder weapon.

The Mañana group did not last long. By the middle of 1943, the authors had been drafted, not for their writing or fighting skills but for their scientific pedigrees. In its dissolution the group provided one more good story. When word got out that

Heinlein, Isaac Asimov, and L. Sprague de Camp had all been sent to work at a research laboratory at the Philadelphia Navy Yard, rumors spread like wildfire among science fiction fans that they had been ordered by the Navy Research Board to create a think tank, heading a project that aimed to make their own futuristic inventions, "super-weapons and atom-powered space ships," into realities. The truth was a little more prosaic; the three had been called up by the Materials Laboratory in Philadelphia but in order to investigate, among other things, hydraulic valves for naval aircraft, "exercises in monotony," as de Camp called them. Science fiction fans faced other disappointments as the war continued. In New York, at the site of the 1939 World's Fair, the 4,000 tons of steel used to construct the Trylon and Perisphere were pulled down and transferred to military factories for scrap metal. Even futurist dreams were being bent to the war effort.

As the prophets of the future were compelled to devote themselves to the problems of the present, so the GALCIT project was also undergoing major changes at the military's insistence. British Intelligence had recently discovered a V-2 missile construction plant at Peenemunde, Germany. Hitler had become convinced that the V-2 would be the decisive factor in Germany's winning the war. His belief was understandable. Thanks to the efforts of Wernher von Braun, the remnants of the VfR, and huge military backing, the V-2 was the most advanced rocket ever constructed. The height of a four-story building, the V-2 was a liquid-fuel rocket, powered by ten tons of liquid oxygen and ethyl alcohol, its huge engine consuming this vast amount in little more than a minute. By that time, however, the rocket had reached an altitude of 17.4 miles (28 kilometers) and was traveling at a speed of 3,500 miles per hour (5,630 kilometers per hour). The V-2 was designed to carry one ton of explosives over a distance of more than 200 kilometers. Beginning in the autumn of 1944, over four thousand were fired at London and Antwerp. The British Intelligence reports were sent to California for Kármán to review with the assistance of Malina and, thanks to Kármán's influence, a happily returned Tsien. They concluded that the United States should begin immediate research into the possibility of making similar missiles.

Though it had seen its number of workers and its funding increase rapidly, the GALCIT project was still following the directives of 1939, limiting its efforts to the development and improvement of aircraft rockets. Now the group was given carte blanche to develop undisguised war machines, guided missiles that could fly 150 miles and deliver a payload of some 1,000 pounds of explosives. The United States Army Ordnance Department gave them $3 million and one year to do it.

In the face of such huge demands, the Air Corps Jet Propulsion Research Project was reorganized and renamed the Jet Propulsion Laboratory better known as the JPL. A new board was established to administer the new project, which was now divided into eleven sections analogous to university departments. The number of workers in the Arroyo leaped from eighty-five in 1943 to nearly four hundred by the war's end. One-third of these workers were trained at Caltech.

As the facility expanded, the original rocketeers become less and less involved. Although Malina was named the first director of the JPL, administrative responsibilities meant he was growing distant from the hands-on work he loved. "I was aware of more and more research activities but in less and less detail," he recalled. Moreover, the weapons he had been so loathe to create were now to be the main product of the project which he, Parsons, and the Suicide Squad had founded. Parsons still harbored dreams of travel to the moon—he was telling members of 1003 how he wanted "to do space work with planes" after the war—but it was becoming clear to the rocketeers that the dollars the military were sinking into rocketry were meant solely for current and future wars. The modesty of the JPL's old-fashioned title—it still refrained from using the word *rocket* because of its science fiction connotations—was all that was left of the innocence and simplicity of the original rocketeers' hopes and dreams. In the words of the technical historian, Dr. Benjamin Zibit, "JPL may have been born in the light of intense scientific research, but [it] was the child of war."

The good news coming from Europe helped mollify any moral concerns the rocketeers might still have felt. By the last months of 1944, the end of the war was in sight, and Parsons was no longer the only scientist in Pasadena who was playing as hard as he worked. When not at Aerojet or visiting the Arroyo Seco, he and others from the original group of rocketeers could often be found at one of the many alcohol-soaked gatherings at Andrew Haley's new house—a mansion located, naturally, on Orange Grove Avenue.

Haley embraced the enchanted atmosphere of Orange Grove just as Parsons had as a child. He held frequent parties at which the scientists, along with whatever actors and actresses, politicians and movie moguls Haley had been doing business with, would drink late into the evening. Despite the fact that war rationing made alcohol harder to get hold of than petrol, Haley and his wife Delphine made sure that their liquor cabinet was stocked, ready to lubricate the minds, wits, and tempers of the rocketeers and profiteers alike. There was nothing Haley liked more than to instigate an argument. "He wouldn't settle down for a dull conversation for a minute," remembered his son, Andy Haley Jr., who was six years old at the time. Sparkling discussions would often ferment into arguments before frothing once more into jokes.

Despite the glitz and glamour his connections afforded him, nothing seduced Haley like talk of outer space. He sincerely admired the rocketeers, and they formed the core of the many parties that took place at his house throughout the year. These strange academic guests heightened the magic of the Haley children's world. Kármán was "a very sweet man," remembered Andy Haley Jr., although he was "a little bit frightening with the bushy eyebrows, he kind of mumbled and talked in vague sentences." Fritz Zwicky would argue in his thick Swiss accent on his pet topic, "How many dimensions does an angel have?" Frank Malina was more serious, concerned with the war in Europe and prone to speculation about possible future conflagrations that were coming their way. Ed Forman usually arrived like a "young dude" on his motorbike, which would be buckled or bent from his latest accident. But in the young Andy's eyes, Parsons' arrival made the deepest impression. "On occasion he would love to terrify us kids, and me especially. He would be wearing this snake, a big cobra around his neck, and he would give it to me to hang up for him, as if it was a coat."

Dede Haley, Andrew Haley's ten-year-old daughter, was not so much terrified as smitten. "Jack Parsons to me was just the living end," she recalled. "Oh, he was just a very handsome man." Nevertheless, she found there was something unmistakably disturbing, even wicked about him. "He had a slight satanic look, he really did, it was something about his eyebrows, in the middle of the eyebrow it would go up in a kind of inverted V." The childrens' overjoyed father would greet Parsons with his usual exhortations, "O god, get in here immediately, oh dear soul." The two shared a love for jokes. On Halloween, Parsons got Kármán, Haley, and a few of the others to dress up as ghosts. "When kids came up to the house they opened the door and scared the kids away. They thought that was great," remembered a friend of Parsons from the time. "They had a kind of a childish sense of humor, just big boys at the time."

Haley was particularly fond of Parsons' ability to declaim reams of poetry from memory. He would often call 1003 in the middle of the night, shouting, "Get over here right now and recite the Ode to Pan!" remembered Jeanne Forman. "Sometimes he would call up at one o clock or two o clock in the morning . . .

or three sometimes. Everybody would get dressed and run like hell to Andrew Haley's to hear Jack recite 'Io Pan! Io Pan!'"

The "Hymn to Pan" had long been one of Parsons' favorite Crowley poems, and he was not shy about sharing it. As the other members of Aerojet milled around on the patio outside Haley's house, arguing and drinking, Haley would usher Parsons up to the balcony that overlooked it. Then, after the Aerojet crew had quietened somewhat, Parsons would begin his recitation, slowly stamping his feet in time with the meter:

> *Thrill with lissome lust of the light,*
> *O man! My man!*
> *Come careering out of the night*
> *Of Pan! Io Pan!*
> *Io Pan! Io Pan! Come over the sea*
> *From Sicily and from Arcady!*
> *Roaming as Bacchus, with fauns and pards*
> *And nymphs and satyrs for thy guards,*
> *On a milk-white ass, come over the sea*
> *To me, to me . . .*

Often by this point the sight of Parsons, ecstatic, sweating, head tilted back to the sky, was too much for the partygoers, and he would have to halt his recitation to beat off projectiles launched at him from the ground. Andrew Haley, tears of laughter streaming down his face, would encourage him to continue. Indeed, Parsons' attempts to finish the poem might be drowned out by the drunken voices of the Aerojet crew as they launched into an equally fantastic song that Andrew Haley's sister had written for them, sung to the tune of the old Irish folk song, "Garry Owen":

> *Oh come ye sons of Aerojet,*
> *The way is clear, the course is set;*
> *The beacon is the vast unknown,*
> *So let us sing the chorus.*

To the sun, the moon, the stars away,
Our Aerojet will reach some day;
And all the world will wondering say,
It's like a fairy story.

The party would continue on until dawn, while the children, thrilled to be part of this strange adult world of fantasy, would sneak over to the Valley Hunt Club and splash until dawn in the swimming pool.

Such performances as these nurtured the long-standing rumors among Aerojet and Caltech staff that Parsons was involved in some form of "mythic love cult" and linked to women of "loose morals." Some knew of 1003 as a "gathering place of perverts." At best Parsons was known as "a character," at worst "a crack-pot."

Rocketry was swiftly becoming more specialized, more like an orthodox science. In 1936 Malina had estimated that there were fewer than fifty engineers in the world seriously interested in astronautics; now there were several hundred. There was no room for a hands-on innovator like Parsons within the new system. Traditionally trained scientists from other fields were now becoming involved, and they expected others to follow their rules. The old familial atmosphere was rapidly disappearing.

Dorothy Lewis, Frank Malina's secretary at the JPL, remembered a distinct change taking place in 1944. "I don't remember that it was the same little group anymore ... It was not as close as when we were in the primitive building and everything was new, and they all had to work together to get anything done ... It became much more formal. And these outside people—you can't blame them, they hadn't been part of the inner circle." The days when Parsons could get by without ever writing up his calculations were over. "For testing we'd plan the work and about ten to fifteen mechanics would do our work for us," recalled Charles Bartley, a research engineer now working on solid fuels at JPL.

But Parsons was still addicted to experimenting; and he still treated both Aerojet company and the JPL's laboratories as his own private domain, often appropriating particularly interesting chemicals for his own recreational explorations. At Aerojet's new test site in Azusa, fifteen miles west of Pasadena, Parsons' habits provoked the ire of Fritz Zwicky, professor of astrophysics at Caltech and a man not used to being disobeyed. Zwicky did not care much for the untrained Parsons, and he later remembered him as "a dangerous man." Calling him "a collector of explosives . . . Like people collect postage stamps, Parsons was enamoured with explosives." Andrew Haley had asked Zwicky to become scientific director of Aerojet, and as such he was Parsons' nominal supervisor in Aerojet's Research Department. Zwicky had become "very excited" about using nitromethane as an oxidizer for liquid rocket fuels, instead of the acid-analine mixture which the Suicide Squad had come up with some years before. Parsons, who had studied nitromethane when Aerojet and JPL were little more than a pipe dream of the Suicide Squad, knew that this was a bad idea. Nitromethane was far too unstable and dangerous to be used in rockets. Zwicky may have been a scientist of greatness, a student of Einstein's no less, but in terms of explosives knowledge, the thirty-year-old Parsons was still his superior.

Not knowing of Parsons' objections, or more likely indifferent to them, Zwicky ordered a batch of the chemical to be delivered to the Aerojet site. When Parsons found out, he must have been upset by Zwicky's refusal to heed his advice. He and Forman sneaked into the site and gleefully exploded the whole batch, "blowing up half the business," according to a furious Zwicky. Zwicky could hardly fire a founder of the company, however; and when the affair came in front of the company board, Malina backed Parsons' stand against the use of nitromethane. "It would have been a disaster for Aerojet if that proposal had been followed," he recalled.

Though Parsons was proved right in the end, he hadn't endeared himself to anyone. His colleagues were also starting to grow weary of his increasing openness about his private interests and his continued dalliances with the company's secretaries. "We told him all the time, I mean, all these fantasies about Zoroaster and about voodoo and so on, this is okay; we do that too in our dreams. But keep it for yourself; don't start impressing this on poor secretaries," remembered Zwicky. "I mean he had a whole club there you know."

The new members of Aerojet and JPL had little time for such eccentricity, and all the while Parsons' behavior was growing more outrageous. Other Aerojet workers could clearly hear him chanting and stamping his feet at rocket test firings "like Billy Graham," although what he was actually chanting was Crowley's pagan "Hymn to Pan."

Parsons' life was growing increasingly hectic as he tried to balance both his personal and professional lives. Writing to Forman, he described the chaos with glee:

> Just a note from the midst of the vast sea of confusion wherein I dwell. Navy training programs, late tax returns, overdue reports to A.C. [Aleister Crowley], and of course Aerojet and its marvelous personalities . . . Production is six months behind on everything, which is O.K. because the engineering releases are two months behind that. (You'd better burn this).

However as the late nights at 1003 caught up with his early morning starts in the Arroyo, Parsons increasingly disheveled appearance led to much sniggering behind his back at the JPL. The research engineer Charles Bartley remembered, "Jack would always use a perfume and wouldn't take a bath so he had a very strange odor about him most of the time." Complaints against him may have been sharpened, too, by professional jealousy. Although officially the Aerojet Solid Fuel Coordinator,

Parsons was working on the problem of smokeless JATOs for two or three days a week at JPL and getting paid double. Even Apollo Smith, one of the original Suicide Squad, griped about it. "They [Parsons, Summerfield and Forman] were able to authorize their own consulting time [at JPL]. And this just didn't seem right to me. It just seemed to me that, well, if they needed a little money, they could come in and consult, whether it was wanted or not."

While the JPL had become a fully funded military concern, Aerojet was transforming itself into a regular business which demanded regular personalities willing to carry out regular jobs. It was a particularly inauspicious time to be a loose cannon: The Navy was now demanding 20,000 JATOs a month. The company had managed 2,000 the previous year, and even then by the skin of their teeth. A continuing lack of capital investment threatened to make the monthly order impossible to fill. The company began seeking outside investors.

Alarmed by the thought of a takeover, Parsons threw himself into the role of the reliable company man. Official letters by him suddenly appeared in the company files, as if he was making a special effort to fit in, to be a regular worker. He sought the advice of the chemist Linus Pauling about further fuel experiments, showing that he could work in harmony with Caltech. He conferred with Malina at JPL about similar problems with fuel they both faced, and he attended meetings more regularly. But he could not shake his old working practices. He still preferred to follow his own hunches rather than work from the results of other scientists. "I am aware that it may be said that this is impracticable, has been investigated, better processes are known etc etc., ad tedium," he wrote about a new process for making liquid fuel. "Nonetheless, I should like to undertake this investigation . . . I have already prepared samples which look encouraging in my home laboratory, and I should like to take about a week to bring it to the test stage." He still conducted himself as a pioneer of the science, rather than as one of many in the field.

Thanks to the salesmanship of Andy Haley, the General Tire
and Rubber Company agreed to buy a majority shareholding in
Aerojet. (General Tire's president had originally been interested
only in acquiring a radio station that Haley was the attorney
for, until Haley sold him on the idea of Aerojet's potential.)
However, it was made clear, from the Caltech contingent within
Aerojet, that if the sale was to take place, Parsons and Forman
should not be included. As Zwicky put it, "You couldn't have a
guy like that associated with a reputable institution." With the
rueful agreement of Frank Malina and Martin Summerfield, it
was left to Andrew Haley, friend as he was to the young men, to
persuade them that they should sell their stocks to the other
shareholders. If they had known about the conditions for the
sale, the two might have hung onto their shares out of stubbornness. But Haley withheld this knowledge from them, and given
the likelihood of Aerojet's profits dropping at the end of the war
and the fact that Parsons still carried the entire financial burden
for Agape Lodge, the benefits of selling out for cash up-front
would not have been too hard for Haley to conjure.

In December 1944, General Tire bought 51 percent of Aerojet's shares for $75,000. Forman sold his stock to Malina for
$11,000. Although no records survive of Parsons' sale, he prob-
ably earned a similar sum. Certainly it was not a bad return for
an initial investment of $250 some three years before, but oth-
ers would go on to make much more. Today Aerojet is a multimillion dollar company employing over 2,500. The company's
propulsion systems have been used on every manned space vehicle the United States has launched, and the company continues
to create solid- and liquid-fuel rocket engines for space agencies
across the globe. By the 1960s Parsons' shares would have been
worth over $12 million. "They were cheated in a way," remem-
bered Zwicky.

Parsons still worked occasionally as a consultant at Aerojet—nobody knew explosives quite like him—but that arrangement would not last long, and his remaining links to the JPL

would disappear with it. Caltech professors like Zwicky acknowledged his work in an offhand, patronizing manner—"He
had done a useful thing by starting these playthings down here
in the Arroyo"—but Parsons the innovator and explorer was no
longer needed. Despite the proceeds from the sale, Parsons was
not entirely blind to his predicament. He confided to his old
Pasadena friend Robert Rypinski that he wanted to go on experimenting and developing at Aerojet, but "they just didn't have
the room for financing his experiments or listening to his ideas."
With the army now beginning to invest heavily in the production
of rocket weapons, many of the aviation companies such as
Northrop, Lockheed, and McDonnell started up rocket research
and development divisions. New rocket companies began to
spring up across the country, and even the burgeoning American
Rocket Society was trying to cash in on the craze they had helped
begin, creating their own company, Reaction Motors, Inc. The
age of the gentleman rocketeer, who dreamed of space travel
and worked independently or in small, autonomous groups, was
over. The death of Robert Goddard, four days before the end of
the war, seemed merely to emphasize this. One member of the
American Rocket Society sadly noted that the days when a scientific breakthrough could take place in the backyard were
gone. "The jig was pretty much up for rocket amateurs."

At the age of thirty, Parsons was cut adrift from the world of
rocketry for the first time in his adult life. It was plain to see that,
like Goddard before him, he was being left behind as the very
science he had helped to create soared up and away from him.

If Parsons harbored any grudges about his removal from Aerojet, he did not make them public. In fact, he seemed to be enjoy-
ing himself tremendously. With the proceeds from the sale of his
Aerojet shares, he bought the lease to 1003 and spent almost all
his time there, in the manner of a man without a care in the
world. His favorite game of the moment was what he called "the
Kayam Kontest." The rules were simple, as he explained to his

friend Grady McMurtry. "Take a small glass of Pernod, and a large glass of sauterne. Drink the Pernod, recite a verse of *The Rubaiyat* (beginning with No. 1), then chase both with the sauterne. Try it. The last one to articulate wins. I usually bog down in Jamahyd and Kaikobad." Since these two names appear in verse nine of *The Rubaiyat of Omar Khayyam,* Parsons clearly possessed considerable talent at both drinking and recitation.

He liked to spend hours playing with toy boats in the bath or lounging about the house with Betty. Indolence set in: "He is all enthusiasm, and drops it, he makes plans that are never carried out, he loses interest," sighed Jane Wolfe. The remaining residents at 1003 were beginning to wonder if Parsons would be able to continue to look after them if he did not find himself a new job. But Parsons was making plans of a sort. With Ed Forman he formed the Ad Astra Engineering Company, its name taken from their childhood motto. They envisioned Ad Astra as an umbrella company for their personal pursuits. Parsons would work on explosives consulting and manufacture, creating an explosives company named the Vulcan Powder Corporation. Forman, who had spent most of the proceeds from the Aerojet sale on an open-cockpit plane, in which he proceeded to learn all manner of daredevil stunts, would try his hand at the latest boom industry, Laundromats.

May 8, 1945, saw victory in Europe. Though the war still raged in the Pacific, somnolence reigned over Parsons' California life. Even the Japanese balloon bombs, incendiary devices drifting slowly across the Pacific and exploding up and down the West Coast, seemed little more than a valediction to war. On the Fourth of July, Parsons treated a group of guests to a spectacle—a six-foot-tall firework which he had built in the house's basement. With a crowd gathered around him in the garden, he launched the rocket with a roar and all watched it soar out across the Arroyo Seco. Unfortunately, the sparks from the rocket leaped into the straw-dry grass, and a fire broke out. Residents and guests alike grabbed buckets of water and splashed and stomped out the fire. When the sound of sirens was heard, the party scattered.

Two days later some deadlier rockets arrived in town: Two V-2 missiles rolled into the Pasadena train station on an open flatbed rail car. Allied forces had recently captured an underground factory in Germany's Harz Mountains, where the weapons had been manufactured by slave labor. They were now at Caltech to be studied. Painted in black and white squares, the forty-six-foot-long missiles had a vicious, sleek grace. These rockets held within them the promises of space and the future. The people of Pasadena thronged to see them; surely Parsons, feeling envy as much as awe, joined the crowd.

In early 1945 Regina Kahl died. The stresses and machinations of the OTO had been cruel to her blood pressure. Parsons assembled the Lodge members at 1003 and raising a wine-filled glass, declared, "There is no part of her that is not of the gods." He drank the wine and shattered the glass in the fireplace. In many ways it was the final nail in the coffin of the old OTO. Many of the old members—including Jane Wolfe, most senior of all—had left the house but not the OTO after the Smith debacle, and Parsons had been turning the Lodge into something closer to a secular bohemian boarding house. Some of the new tenants were acquaintances from the LASFS. The science fiction artist Louis Goldstone, fan Alva Rogers, and the up-and-coming journalist Nieson Himmel had all heard stories of Parsons' rocketry background and occult excesses. He was just the sort of landlord they were looking for.

Himmel was a "Buddha-like figure," five feet seven inches tall and on his way to weighing nearly three hundred pounds. Only twenty-three years old, within two years he would become one of Los Angeles' most prominent crime reporters, covering the infamous Black Dahlia murder and the assassination of Bugsy Siegel. Goldstone, who had been introduced to Parsons by McMurtry, wrote to McMurtry describing the move. "Since being around here a month has got me out of the apologizing habit, I'll not present you with any . . . I needn't tell you how happy I am here with Jack and Betty and this wonderful place. I'm attending classes at the Art Centre in Los Angeles and, I think, doing pretty well. In addition I'm leading a private life amid the best company and surroundings I've yet known—a life more vital and interesting than I could have anticipated before I came here. Indeed, this wilderness is paradise now." For Alva Rogers, nicknamed 'Red' for the color of his hair and his politics, meeting Parsons was quite a shock. "Jack was the antithesis of the common image of the Black Magician one encounters in history or fiction . . . He was a good looking man in his early or mid-thirties, urbane and sophisticated, and possessed a fine sense of humour. He never as far as I ever saw, indulged in any of the public scatological crudities which characterized Crowley," although "he did have the Crowley approved attitude toward sex."

A rather different tenant was Robert Cornog, six feet five inches tall and athletically built. Before the war he had held an assistant professorship at Berkeley, where he had discovered the radioactive isotope, tritium. Now he worked as chief engineer of the Manhattan Project's ordnance division and was intimately involved in the development of the trigger mechanism for the atomic bomb. His coworkers saw him as a "rugged individualist, frank and outspoken," as well as slightly eccentric. He was renowned at the project for his intense interest in physical fitness and could often be seen picking up large rocks and tossing them around for exercise. He met Parsons through their mutual friend, Robert Heinlein, whom Cornog originally knew from a nudist colony in Denver in the 1930s. Now that Cornog's work on the Manhattan Project required him to travel between Los Alamos and Caltech, Heinlein put him in touch with Parsons, who still had spare rooms available. The two got on well. Cornog had been an early member of the American Interplanetary Society, and he shared a love of rocketry with both Parsons and Heinlein. There were unmistakable parallels between Cornog and Parsons: Both were eccentric scientists, interested in left-wing causes, working on groundbreaking projects. Cornog came to stay more and more frequently at 1003 over the coming months.

Parsons, who needed to rent out every room in the house to make ends meet, now placed advertisements in the local newspaper. Alva Rogers remembered them well: "Jack specified that only bohemians, artists, musicians, atheists, anarchists, or other exotic types need apply for rooms—any mundane soul would be ceremoniously ejected. This ad, needless to say, created quite a flap in Pasadena when it appeared."

Because of an extreme housing shortage in the country, many people applied, and Parsons and Betty handpicked those they thought would best fit in the house. "The replies to the ad for boarders was fantastic," remembered Himmel; "I was already in the house at the time. People were all insisting how they were atheists."

"Those accepted were a typically peculiar bunch," remembered Alva Rogers. "The professional fortune teller and seer who always wore appropriate dresses and decorated her apartment with symbols and artifacts of arcane lore; a lady, well past middle age but still strikingly beautiful, who claimed to have been at various times the mistress of half the famous men of France; a man who had been a renowned organist for most of the great movie palaces of the silent era." Occultists rubbed shoulders with nuclear scientists. Rocket scientists ate breakfast with science fiction aficionados. Children ran freely through the house.

When the recently released convict Paul Seckler returned, 1003 housed possibly the strangest collection of people ever to live under one roof. Those who were not OTO members would often see Parsons leading the group in black and white robes towards the pergola, where they performed the Gnostic Mass. He had managed to get one of the older members, Roy Leffingwell, to compose some new music for it. Betty, of course, played the role of the priestess. Ed Forman's stepdaughter Jeanne Ottinger, remembered sleeping over at the house. "I can remember one time waking up in the middle of the night to this chanting sound and mother wasn't in the room with me and so I got up and I opened the door and peeked out and I saw they all had these long black robes on and they were chanting." When her mother returned and saw her daughter's querying eyes, she was quick to dispel any worries. "They're having a party," she told her.

By the end of the summer, there were some twenty people living in the house and coach house. Rents were negotiable. "I probably paid $30 [a month]," remembered Himmel. "Alva never seemed to have money for rent, and the girls almost certainly never paid." Parsons' friend Robert Rypinski, home on leave from wartime duty in the navy, remembered visiting Parsons at about this time. "In the living room up on the second floor we sat and talked while couples kept strolling in and out, [and] were sitting across the room necking while we were talking (he didn't notice it so I thought I wasn't supposed to)." Relationships might have been loose and free at the house, but it was clear that everyone was in love with one woman: Betty. "She was young, blonde, very attractive, full of *joie de vivre,* thoughtful, humorous, generous, and all that," remembered Rogers. Himmel was similarly infatuated. "The most gorgeous, intelligent, sweet, wonderful person. I was so much in love with her, but I knew she was a woman I could never have." She "wasn't sophisticated but a natural beauty . . . Sometimes it was hard to look at her directly in the eyes."

When the group went swimming at the Pasadena Athletic Club (where Betty's father was a member), many of the male members of 1003 could do little but gawk at her. "I can still describe Betty's swimsuit detail for detail," recalled Cornog fifty years later. Still only twenty-one years old, she was meant to be taking a writing course at UCLA, but more often than not she skipped her classes and stayed at 1003. Although Parsons owned the house, it was Betty who ran it. "She was pretty skill-ful at running the household. She was the one that bought the food, we'd give her our stamps, our food stamps and she'd buy the food and cook it and serve it. And she ran the place pretty much with an iron hand," explained Alice, Robert Cornog's wife. She also took to inviting her friends to stay at the house and to sleep wherever there was room, much to the older OTO members' displeasure, who, when attending the mass, turned their noses up at these "pseudo-bohemians."

Despite the couple's frequent affairs, Parsons and Betty's rela-tionship seemed to Rogers "permanent and shatterproof." In Betty, Parsons had a companion who could charm and flirt and also act as his partner in magick rituals. Some members of the OTO began to grow concerned about his fixation on her. There were worries, expressed in the usual flurry of backbiting letter writing, that Parsons was falling, as he had with Smith, under the spell of a dominant personality. But Betty was not a woman who was easily awed. In the past she had written letters to Crowley chastising him for the constant intrusions of Crowley's spy Max Schneider. "Can you imagine a dinner party with scientists from the Institute and Max at the table, pompous little Max, with his complete lack of subtlety and humor," she had complained.

Fred Gwynn, a new OTO recruit living at 1003, was aghast at the control she had over Parsons and thus the Order itself. "Betty went to almost fantastical lengths to disrupt the meetings that Jack did get together. If she could not break it up by making social engagements with key personnel she, and her gang, would go out to a bar and keep calling in asking for certain people to come to the telephone." Smith may have had his faults, some began to grumble, but at least the Order had been at the center of his life. They suspected Betty wanted nothing more than to do away with the OTO altogether. Germer labeled her "an ordeal set by the gods" for Parsons. He was not entirely wrong.

At the moment, however, Parsons was concerned about something else: the return of his old mentor, Wilfred Smith.

Smith had removed himself, as per Crowley's instructions, to Roy Leffingwell's turkey farm, a hundred miles to the northeast of Pasadena in the desert town of Barstow; but he had not discovered his true divine nature. Now with no money and with fewer prospects, he returned to 1003. Helen was supporting the family as a munitions worker at the Hughes Aircraft Company.

Smith still held an allure for Parsons, who wrote to him, "I shall always honor, revere, and love you as my teacher, my guide," and continued to give him money. But he also wrote to Crowley, "Smith is a menace." Parsons worried that since he and Helen were still officially married, he might become responsible for Smith's son, Kwen. Although both Smith and Helen had assured him that this would not be the case, Parsons wanted a formal divorce. Helen was startled by this demand, which she saw as hopelessly bound to "old Aeon" rules, but Parsons insisted. He agreed to recompense her for the money and effort she had devoted to him in his early rocketry days, offering her $500 in cash, 10 percent of any income gained from the Ad Astra Engineering Company, as well as $2,000 if the house on Orange Grove was sold. He also installed her and Smith in 1003's coach house until they could find other lodging.

The lack of rocketry work, the divorce proceedings, and the problems Smith's presence caused between him and Crowley began to undermine Parsons' usual indefatigable confidence. When he visited Jane Wolfe, he was visibly disturbed, hardly the carefree man he had been when he left Aerojet nine months previously. In a letter to Crowley, she wrote, "Last evening, when Jack brought me these various papers to post to you, I saw, for the first time, the small boy, or child. This it is that is bewildered, does not quite know when to take hold in this matter, or where, and is completely bowled over by the ruthlessness of Smith— Smith who has a master hand when it comes to this boy."

Following a close call with a German bomb, Crowley had moved out of London to a boarding house in Hastings called

Netherwood, a large Victorian building standing in quiet, wooded grounds. By now illness and addiction had thoroughly emaciated him and his old suits flapped loosely from his gaunt frame. He had lost little of his perspicuity however. When he heard of Smith's reemergence, he was as exasperated as always. "Have I got to explain to everybody all over again, that from the point of view of the Order and of every member of it, Smith is *dead?* The decision is irrevocable." As for Helen, he declared, "If she intends to stick to Smith, I think she ought to be *suspended from the Order while that condition remains.*"

But Crowley knew that even his most fearsome letters could not prevent Parsons from succumbing to Smith's immediate presence while Smith stayed at 1003. "Jack is a bit like a marshmallow sundae," he wrote rather confusingly to Jane Wolfe. "He does what the last person to talk to him tells him . . . He is, moreover, too ready to emphasise the sexual side of life . . . Science, art, philosophy and the like are our prime care . . . We must intensify, concentrate, exalt this side of our nature. *Do* get Jack to see this! He has so much A.1. in him."

Crowley wished he could persuade Parsons to visit England, "for six months—even three, with a hustle—to train in Will, in discipline." But already he was shifting his attentions to another disciple, Parsons' protégé and friend, Grady McMurtry, who because of his military service in Europe had been able to visit Crowley in London and now in Netherwood. He played chess with the Great Beast and learned magick from the very fount of arcane knowledge. If Parsons could not tear himself away from Smith and Betty, and restore order to the OTO in California, then McMurtry was shaping up to be an ideal replacement.

On July 16, 1945, at 5:30 A.M. in the desert of New Mexico, the first atomic bomb test—named Trinity—took place. A bulging boil of fire exploded outwards with the force of some 18,600 tons of TNT before swelling into an imperious mushroom cloud. It melted the desert sand into a glassy olive-green substance in which the scientist was seen transformed from a man of learning into a jadelike god of destruction. Quoting Vishnu in the Bhagavad Gita, Dr. Robert Oppenheimer declared his apotheosis, "Now I am become Death, destroyer of worlds." On August 6, less than one month later, the bomb was dropped on Hiroshima.

In Parsons' circle reactions were varied. Crowley wrote excitedly to one of his disciples, explaining that *The Book of the Law* had foretold the bomb in its verses, "I will give you a war-engine / With it ye shall smite the peoples; and none shall stand before you." The science fiction fans were stunned by, and almost perversely proud of, the prescience of their beloved stories. As John W. Campbell wrote in that month's editorial in *Astounding Science Fiction*, "The science-fictioneers were suddenly recognized by their neighbors as not quite such wild-eyed dreamers as they had been thought, and in many soul-satisfying cases became the neighborhood experts." It was left to the scientist Malina, still working at the Jet Propulsion Laboratory, to express pessimism and confusion, writing, "Atomic energy is here and no one knows for sure what it means, except that another war might be worse than the last one—if that is possible."

Two months after the blast, Malina had the chance to fly over the Trinity test site. He found it "a very disturbing sight, especially for us who were involved in the development of long-range rocket missiles." But his concerns were set to one side as he was traveling to New Mexico in order to make his own mark in history. On October 11, 1945, the WAC Corporal, a sixteen-foot-tall rocket, little more than a foot in diameter and weighing 655 pounds, was launched into the desert skies. Powered by both solid- and liquid-fuels made at Aerojet, it was a sounding rocket, intended not for military purposes but for gaining meteorological information about the upper atmosphere. The WAC Corporal thus fulfilled Parsons, Forman, and Malina's original Caltech proposal, begun some ten years earlier. Roaring into the blue, it reached a height of seventy kilometers (forty-three miles) and was airborne for some seven and a half minutes. It was the first American rocket to reach such heights. For Malina the launch was both a great and sobering achievement. "I was the sole member of the original GALCIT Rocket Research Group of 1936 to experience the culmination of its hopes after the many vicissitudes in rocket development over the ensuing period of 10 years," he wrote. "It is difficult to describe my feelings as I watched the sounding rocket soar upwards. One can think of many things in a few minutes. One of my thoughts was that I could now turn my mind to other goals in a world full of both fascinating technological possibilities and of desperate social problems."

Within less than one year Malina would leave the Jet Propulsion Laboratory, the last of the Suicide Squad to do so. But the Squad's influence, however unheralded, would resound through the years. They had established a firm theoretical and experimental ground for rocketry and had helped make it a valuable science in the eyes of the government and the military. They had prompted Caltech to offer the nation's first academic course in rocketry, and they had created the Jet Propulsion Laboratory, the institution that was to become America's first center for long-range missile development and space research, producing, among many other spacecraft, *Mariner 2,* the first craft to orbit another planet, in 1962, and *Viking 1,* the first craft to land on Mars, in 1976. In short, the Suicide Squad's romantic idealism ushered in a revolution. They opened a window of opportunity for future scientists, triggering a movement that would propel the United States into space and towards the moon.

In 1949 the JPL mounted a WAC Corporal onto the nose of a captured V-2 missile. The combination, known as the Bumper-WAC, was symbolic of the mix of homegrown and foreign know-how that would fuel the United States' space program for the next half-century. When the V-2 reached its maximum altitude, the WAC Corporal's engine was ignited, allowing it to reach a height of 244 miles (393 kilometers). The launch was man's first step into space. A new age was born.

11

....

Rock Bottom

*I shall attempt to evoke the true image of one who assumed
with plausibility in an age of science the long-discarded robes
of prophecy.*

—Edmond and Jules de Goncourt, *Journal*

*The best lack all conviction, while the worst / Are full of
passionate intensity.*

—W. B. Yeats, "The Second Coming"

At the end of the war, many of Parsons' friends and colleagues
returned to California and Pasadena from their military post-
ings. The nuclear physicist Robert Cornog now moved into the
house full time with his young family, and Robert Heinlein, recently returned from his engineering work in Philadelphia,
would stop by occasionally to visit his friend. Anthony Boucher,
Edmond Hamilton, and Jack Williamson all stopped by for
short visits, too, creating a state of almost religious ecstasy
among the LASFS contingent living there. Thus, no one was surprised when Lieutenant L. Ron Hubbard, another fantasy writer
of great renown, announced his plans to move into 1003.

Three years older than Parsons, Hubbard had red hair, a
wide mouth, and a plump face, lent definition by horn-rimmed
glasses. At the behest of LASFS member Lou Goldstone, Hubbard had made a brief visit to 1003 earlier in the year, and Parsons had extended an open invitation to return. Hubbard had

burst onto the science fiction scene in 1938, having previously written westerns and sea stories. His tales, which appeared in *Unknown* magazine, became particular favorites of Parsons over the years. They often attributed exceptional powers to the mind, envisioning the power to heal or kill by thought alone. Before the war, Hubbard had been an occasional visitor to Robert Heinlein's Mañana Literary Society, and it is possible that he first met Parsons there. Even if they did not meet in fact, they definitely met in fiction, for Hubbard also appeared, thinly disguised, in Anthony Boucher's *Rocket to the Morgue*. As Jack Williamson noted in his memoirs, Hubbard was the model for the charismatic seducer and prime murder suspect, Vance D. Wimpole.

From the interviews and memoirs of the other residents at 1003, it seems clear that Hubbard had a personality that either charmed or repulsed immediately. Born in 1911, he had managed to fit a lifetime's worth of adventure into the intervening years. The breadth of his accomplishments seemed to rival those of the other polymath in Parsons' life, Aleister Crowley. Upon his arrival at 1003, Hubbard dominated conversations with his anecdotes and wit. When he spoke at the communal supper table, no one could steal the floor from him. He claimed to have been involved in counterintelligence during the war and told of escaping from Japanese-occupied Java on a raft with bullet wounds and broken feet. He said he had gone on to command an antisubmarine escort vessel in the Atlantic, although not with great success, for it had been sunk four times. However, once reassigned to the Pacific, his depth charges sank two Japanese submarines. He had managed to embroil himself in adventure even while patrolling the frozen Aleutians, lassoing a polar bear on an ice floe, which in turn climbed aboard the boat and chased the crew, and him, off it.

His peacetime stories were equally extravagant. He was a member of the famed Explorers' Club. Alva Rogers, writing of his time at 1003 for the fanzine *Lighthouse,* remembered

Hubbard proudly displaying the scars he had received after being struck by "aboriginal arrows" on one expedition he led. He had written the screenplay for the 1938 Hollywood film *The Secret of Treasure Island,* based on his novel *Murder at Pirate Castle,* and he had also worked as a "balladeer" for a Washington, D.C., radio station, performing his own works on air while accompanying himself on the ukulele. It was little wonder, the residents of 1003 concurred, that the scientists at the British Museum who measured his skull declared it to be unique.

For the most part the housemates were delighted with their new guest's stories and his back-garden heroics, but a few failed to succumb to Hubbard's magnetism. Jack Williamson had heard Hubbard's naval tales before at meetings of the Mañana Literary Society. "I recall his eyes, the wary, light-blue eyes that I somehow associate with the gunmen of the old West, watching me sharply as he talked as if to see how much I believed," said Williamson. "Not much." Nieson Himmel enjoyed pointing out discrepancies in his stories, much to Hubbard's irritation. "I can't stand phoneys and he was so obviously a phoney. But he was not a dummy. He could charm the shit out of anybody." Alice Cornog put her feelings more simply: "I thought he was a bastard, I disliked him thoroughly." There was no denying, however, that, love him or hate him, believe him or doubt him, Hubbard "told one hell of a good story."

It was not particularly surprising that the man who should fall most thoroughly under Hubbard's spell was Parsons. Enamored of Hubbard's life of adventure, he offered him a bed in the main building, sharing a room with the gadfly Himmel. Soon Hubbard was absorbed quite happily in the house's activities. When he was not writing, he spent much time in the company of Parsons, who excitedly explained to him the laws and sanctions of Thelema. Hubbard impressed Parsons with his immediate understanding of Crowley's work and with his insight—most likely garnered through his years of fantasy writing—into magic

in general. Parsons wrote excitedly to Crowley to tell him about
his new friend and prospect.

> About 3 months ago I met Capt L Ron Hubbard, a writer
> and explorer of whom I had known for some time. He is a
> gentleman, red hair, green eyes, honest and intelligent and
> we have become great friends ... Although he has no formal
> training in Magick he has an extraordinary amount of ex-
> perience and understanding in the field ... He is the most
> Thelemic person I have ever met and is in complete accord
> with our own principles.

Hubbard and Parsons liked to fence together in the large living room, perhaps occasionally competing with Heinlein, who
was also a keen swordsman. When Heinlein visited 1003 to talk
to Robert Cornog, Hubbard and Parsons would join them for
discussions on science and science fiction, batting ideas back
and forth between each other. Before long, however, the house
residents realized that Hubbard's magnetism extended well beyond the genial. "He was irresistible to women, swept girls off
their feet. There were other girls living there with guys and he
went through them one by one," remembered Himmel.

Hubbard had been at the house for little more than two
months when Grady McMurtry arrived back from Europe, fresh
from the war and from studying at the hoof of the Great Beast.
Crowley and Germer had asked him to write a report on the activities of the Agape Lodge based on interviews with every OTO
member. In the report, still in the OTO archives, McMurtry
bears unwitting witness to the early stages of the disruption
Hubbard would inflict on Parsons.

He described watching Parsons and Hubbard fence against
each other one evening, as usual not wearing masks. "The light
was very poor and they kept tangling with the rugs but, as both
men know something of the sport, it was not exactly mortal

combat." Betty was watching from the sidelines, eyeing Hubbard and growing more restless by the minute. When Parsons offered her the chance to try her hand with his foil, she snatched at it and launched into a wild attack on Hubbard, lunging dangerously at Hubbard's unprotected face with a fierceness that shocked the watching McMurtry, who "thought someone was going to get killed." Hubbard, regaining his composure after the initial ferocity of the attack, fought the formidable Betty back a few steps and stopped the assault by rapping her smartly across the nose with his foil.

It seemed more like foreplay than fun and games. And in fact, Hubbard "soon fastened on to Betty," remembered Himmel. Bob Cornog remembered stumbling into Parsons' room one morning to find Hubbard and Betty entwined, "like a starfish on a clam." Now all eyes turned to Parsons, whose devotion to Betty had been absolute, to see how he would react.

In the manner of a true follower of Thelema, Parsons had always prided himself on his ability to renounce jealousy. One needed only to look at the remnants of his first marriage to see how successfully he had managed this. Up until now, he had been quite comfortable with Betty's amorous adventures, always confident that she would come back to him in the end. But he was disturbed by the intensity of her relationship with Hubbard. This was no mere dalliance. Parsons put on a grand show of remaining friends with her. They hugged and talked as before, and he accompanied Betty and Hubbard on trips as "the genial elder brother," but when nighttime came, Parsons was excluded from her bed. For Alva Rogers, as for the rest of the house, it was obvious that "Jack was feeling the pangs of a hitherto unfelt passion, jealousy."

There were other houseguests who were more than willing to fill Betty's place, but Parsons was enveloped in emotions which he could not, for once, conquer. The loss of Betty was not made easier by the loose sexual lifestyle Parsons had encouraged at the lodge. To the shock of the non-OTO members, soon Hubbard was "making out with her right in front of Parsons." Group meals were no longer quite the frivolous affairs they had once been. "The hostility between Hubbard and Parsons was tangible," remembered Himmel.

Driven to distraction by the loss of Betty and without the incessant work of wartime rocket research to absorb his attention, Parsons threw himself into the only other part of his life he could control: his magic. He had been turning his interests to the more perverse branches of occultism in his quest to conjure up actual spiritual phenomena. As far back as 1943, Crowley had warned Parsons, "I don't like at all what you say about witch-craft. All this black magic stuff is 75% nonsense and the rest plain dirt. There is not even any point to it." Black magic was traditionally recognized as a form of ritual magic practiced for evil or harmful purposes. Although the press frequently accused Crowley of being a "black magician," he envisioned his magick as a discipline designed to aid in the individual's mental and mystical development. Ever since Parsons had been a boy, however, the dark side of magic had captivated him. "I know that witchcraft is mostly nonsense, except where it is a blind," he wrote to Crowley in 1943, "but I am so nauseated by Christian and Theosophical guff about the 'good and the true' that I prefer the appearance of evil to that of good."

Parsons was now forgoing his OTO colleagues and enacting rituals with his old friend, Ed Forman. Despite his doubts about the reality of Crowley's magick, Forman was always willing to help out with what he saw as Parsons' hobby. As with the pair's scientific work, their occult methods leaned toward the unorthodox. "They thought, 'Let's work on the heavier stuff at the end of the magic book without doing any of the simpler stuff'," remembered Forman's wife, Jeanne. "They were tinkering with magic spells as they had with their rockets." On one such occasion their frivolousness had such a dramatic and unsettling psychological effect on Ed Forman that his family still discusses the story to this day. It seems that Forman was returning to his bedroom late one night following the performance of a ritual, when he felt the whole house shake. At the same time he heard a piercing scream coming from outside his window and looking out of it, he would recall, he saw a number of horrible entities floating outside his window, what he recognized as banshees—female spirits whose wailing warns of a death in the house. With the sound of their screams filling his ears he rushed downstairs to ask the other members of the house if they, too, could hear it, but nobody could. "Up until then he had not believed in Jack's hobby," remembered Jeanne. "Now he was absolutely terrified." The events of that night would unsettle Forman for the rest of his life.

Forman was not the only one suffering from Parsons' devil-may-care attitude. A worried Jane Wolfe now wrote to Karl Germer about Parsons' new pursuits. "There is something strange going on," she said. "Our own Jack is enamored of witchcraft, the houmfort, voodoo. From the start he always wanted to evoke something—no matter what, I am inclined to think, so long as he got a result." Parsons claimed to be impregnating statuettes with "a vital force" by magical invocation and then selling them, leading many of the OTO to worry about the demonic forces he might unleash upon 1003. A form of group hysteria suddenly gripped the OTO members in the house, and they began performing "banishing rituals"—those intended to clear the psychic atmosphere—on a regular basis. Meeka Aldrich, an OTO member who had recently moved into the house, believed that something "alien and inimical" lurked in the house's wood paneling. Others sensed the presence of "troublesome spirits," especially on the third floor.

In recent meetings of the Agape Lodge, Parsons had read aloud from British author William Bolitho's book *Twelve Against the Gods*. It was a metaphor-laden treatise on the "adventurer," the person—such as Alexander the Great, Casanova, or the prophet Mohammed—who sets himself on an unalterable path to a grand destiny. "The adventurer is an individualist and an egotist, a truant from obligations. His road is solitary, there is no room for company on it. What he does, he does for himself ... The adventurer is within us, and he contests for our favour with the social man we are obliged to be."

Parsons now embarked on an adventure that he hoped would allow him to enter the ranks of Bolitho's heroes. He planned a series of magical rituals—a magical working—more ambitious than any he had attempted before. He would later call it the defining work in his life.

Alva Rogers remembered being awakened on a bleak morning in December "by some weird and disturbing noises seemingly coming from Jack's room which sounded for all the world as though someone were dying or at the very least were deathly ill." He went on:

We went out in to the hall to investigate the source of the noises and found that they came from Jack's partially open door. Perhaps we should have turned around and gone back to the bed at this point, but we didn't. The noise—which by this time, we could tell was a sort of chant—drew us inexorably to the door which we pushed open a little further in order to better see what was going on. What we saw I'll never forget, although I find it hard to describe in any detail. The room, in which I had been before, was decorated in a manner typical to an occultist's lair, with all the symbols and appurtenances essential to the proper practice of black magic. It was dimly lit and smoky from a pungent incense; Jack was draped in a black robe and stood with his back to us, his arms outstretched, in the center of a pentagram before some sort of an altar affair on which several indistinguishable items stood. His voice—which was not actually very loud—rose and fell in a rhythmic chant of gibberish which was delivered with such passionate intensity that its meaning was frighteningly obvious. After this brief and uninvited glimpse into the blackest and most secret center of a tortured man's soul, we quietly withdrew and returned to our room.

Rogers believed that Parsons was trying to conjure a demon to dispatch Hubbard, his rival in love. He was not completely wrong. Parsons *was* trying to conjure up a magical being, not to avenge the loss of Betty, but to replace her. "In December 1945," wrote Parsons later, "I performed certain operations to obtain an Elemental mate."

Parsons generally kept poor records of his life, but he documented the next three months in extravagant detail. He was to keep comprehensive records of each day's rituals, rather as if he were working on a rocket experiment. But instead of straining to catch gauge readings, Parsons would be attuning his mind to less predictable phenomena. In order to summon his elemental, Parsons used an old magical system known as Enochian Magic that had been devised by Dr. John Dee, the royal astrologer to Elizabeth I, and that came highly recommended by Crowley. Dressed in robes and possibly under the influence of a narcotic, Parsons carefully consecrated one of his daggers for an operation as complicated and dangerous in his mind as mixing rocket fuel or firing a rocket. He began by tracing five pointed stars in the air and reciting ancient invocations in both English and Enochian, the strange singsong language that Dee thought was passed down to him from angels. Strewn around him on the floor were paper "tablets" covered with arcane symbols and languages. He recited line after line of ominous and obscure scripture: "Dear Thou Me, for I am the Angel of Paphro Osorronophris; this is Thy True Name, handed down to the Prophets of Ishrael." The ritual called for focused masturbation—what *The Golden Bough* would have recognized as "sympathetic magic"—as Parsons tried to "fertilize" the magical tablets around him and bring his elemental to life. Finally, he performed the banishing rites, a symbolic erasing of all the previous rituals, by retracing in reverse all the pentagrams and hexagrams he had drawn in the air. The entire process took some two hours, and it invariably left Parsons both physically and mentally exhausted.

He performed the ritual repeatedly over the following week, so thirsty for results that soon he was invoking his elemental twice each day. He liked to play Prokofiev's Second Violin Concerto on his gramophone as an accompaniment. He started using his own blood instead of semen, and, his mind sensitive with anticipation, he began to record any phenomena he thought were related to his actions. A violent windstorm followed the first ritual. On another night he was awakened by a series of loud and rapid knocks, and a table lamp across from his bed was thrown violently to the floor, even though the windstorm had died out. In a letter to Crowley he announced, "I have been extremely careful and conscientious in this ritual, lending all my will and scientific training to its precision and preparation. Yet nothing seems to have happened ... The wind storm is very interesting but that's not what I asked for."

A few days later he performed the ritual twice more and observed further unexplained phenomena. At around 9:00 P.M. the electricity failed, and Hubbard, who was in the kitchen at the time, called out to Parsons, saying he had been "struck strongly on the right shoulder." Parsons hurried to Hubbard's side and, in Parsons' words, the two observed "a brownish yellow light about seven feet high in the kitchen." Parsons ran back to his bedroom, grabbed one of the swords hanging from the wall, and hurried back to the kitchen to perform a banishing ritual, burning sulphur and tobacco to aid him in his task. The figure "seemed to diminish" and Parsons followed it into the library, all the while drawing pentagrams in the air with his sword. Finally, the form dwindled into nothing. Parsons diligently noted that Hubbard's right arm remained paralyzed for the rest of the night.

Parsons was already impressed by Hubbard's grasp of Crowley's teachings, but now he became convinced of Hubbard's supreme magical sensitivity. The distress Parsons felt about Hubbard's relationship with Betty was quickly overcome by his remarkable will to believe. He invited Hubbard to sit in on his future rituals and the visions continued to appear. On one occasion Hubbard said that he saw an apparition of Wilfred Smith materialize behind Parsons; before Parsons knew what was happening, Hubbard "pinned it to the door with four throwing knives, with which he [was] expert." Later, in his room, Parsons heard the strange raps again and "a buzzing metallic voice crying, 'Let me go free'."

Were these really spirits speaking to him, conjured up by his rituals, or was someone else providing Parsons with the phenomena he so desperately wanted? Hubbard had taken over Smith's old bedroom across the hall, so he and Betty would have been ideally positioned to stage "supernatural occurrences," especially with the help of the house's secret passageways. If Hubbard was using his imaginative skills to give Parsons satisfaction, then Parsons was all too willing to accept the gift.

However, Parsons did not care to speculate about any underhanded tactics. Instead he wrote in his record, "I felt a great pressure and tension in the house that night, which was also noticed by the other occupants." The tension might, of course, just as easily have been caused by Parsons' insistence on burning sulphur in the kitchen and swinging a sword through the living room, not to mention the twice daily rituals chanted loudly in the Enochian tongue.

After two weeks Parsons and Hubbard traveled into the Mojave Desert together. It had been some time since Parsons had come out here for his rocket experiments. More recently he had escaped to the desert for seclusion in which to meditate and practice his magic. His favorite place was marked by the intersection of two massive power lines, their source and goal lost in both horizons. The sagging cables emitted an ominous, enveloping drone, as if they were the antennae of an almighty cicada buried deep beneath the ground. At sunset the two men stood beneath them and suddenly Parsons felt the tension snap. "I turned to him and said 'It is done', in absolute certainty that the operation was accomplished. I returned home, and found a young woman answering the requirements waiting for me." Parsons had summoned his elemental.

Little did Marjorie Cameron know what she was getting herself into when she arrived at 1003 on the January 18, 1946. Like everybody in Pasadena, she had heard the stories about what went on in the house, but they had not disturbed her. Twenty-three years old, she was known as Candy; she stood five feet five inches tall and had a fair, freckled complexion. Her slant blue eyes, "giant red lips" and burning red hair made her the center of attention wherever she went. Born in Belle Plain, Iowa, her childhood had been far from innocent; she told stories of train hopping and midnight trysts. Robert Cornog remembered her unsettling reminiscences: "Girls had crushes on her and had committed suicide [over her] in her home town." During the war she enlisted as a Wave, a woman serving in the United States Navy, and was assigned to the map room of the Navy Chief of Staff in Washington, D.C. She soon went AWOL and was discharged from the service. She had since returned to her family home in Pasadena where her father and brothers worked as electricians and mechanics at Caltech.

Candy was now living as an artist, supplementing her unemployment benefits by drawing fashion illustrations for ladies' magazines. She had visited 1003 the previous year in the company of one of the house's other residents and caught a momentary glimpse of the house's "mad scientist" owner. The two had not spoken, but in the days after her visit, Parsons had asked after her. By January she had been persuaded to return, following a chance encounter with one of 1003's residents. She was enticed by Parsons' interest in her and also by the rumors of intrigue that she was now informed of: "All the things that were going on with he and Hubbard, and the war that was on with Smith . . . I just couldn't wait to get there."

Her arrival coincided exactly with Parsons' return from the Mojave. He was overwhelmed by the magnetism between the two of them. In a letter to Crowley, Parsons declared, "I have my elemental!" describing her as "fiery and subtle, determined and obstinate, sincere and perverse, with extraordinary personality and intelligence."

He immediately began performing sex magick rituals with her, a strange but not atypical welcome for a new member of the household. "I am to invoke continually, this now being possible and easy," he wrote to Crowley. As for Candy she cared little about the mystical dimension of their bedroom antics. "I didn't know very much [about] Jack's magical work. In fact, I probably derided it," she recalled. Nevertheless, for the next two weeks the couple barely left Parsons' room.

The "magical working" he now began was his most ambitious yet. He believed he could incarnate an actual goddess on earth, a female messiah named Babalon. The goddess Babalon (its spelling "corrected" by Crowley from *Babylon* to provide it with a more auspicious cabbalistic number) first appears as a character in literature in the Revelation, where she is described as a scarlet woman riding on the back of the Great Beast. Parsons believed that his Babalon would also ride on the back of the Beast—Crowley—and augment Crowley's teachings. He hoped that this "Babalon Working" would resound his own name through the ages like the name of a William Bolitho hero.

His companions in the OTO had little idea of what Parsons was trying to achieve. Jane Wolfe, writing to Crowley, admitted his magical work was "much too personal for me, and beyond most of my actual knowledge." Crowley was intrigued by Parsons' undertaking, but he also wondered about his pupil's extreme enthusiasms. "It seems to me that there is a danger of your sensitiveness upsetting your balance. Any experience that comes your way you have a tendency to over-estimate. The first fine careless rapture wears off in a month or so, and some other experience comes along and carries you off on its back. Meanwhile you have neglected and bewildered those who are dependent on you, either from above or from below . . . At the same time, you being sensitive as you are, it behoves you to be more on your guard than would be the case with the majority of people."

Parsons paid no heed to this warning. He immersed himself deeper and deeper in his magic with an excitement that bordered on mania. His letters and notes from the time reveal his exaggerated self-esteem, racing thoughts, persistent agitation, and, in the case of Hubbard's visions, poor judgment—all could have been signs of some form of manic episode. His letters to Crowley came at a more furious pace than ever before, and they were full of feverish, staccato sentences and biblical exaltation. "Thrice blessed, I stand beyond pity or passion, my heart in the light, my eyes turned to the highest. Glory, I cry, Glory unto the Beast and unto Babalon, and Hail to the Crowned and Conquering CHILD."

When Cameron embarked on a brief trip to New York, Parsons went out into the desert once more. There he heard a voice speaking to him, dictating to him just as a spirit had dictated Crowley's own *Book of the Law*. Sitting in the desert, Parsons began writing down a long list of lamentations, declarations, and ritual instructions to form what he called the *Book of Babalon*. Parsons intended his divinely inspired book to be a fourth additional part to Crowley's own volume. But Parsons' book is a jumble of archaisms and colloquialisms, its sole coherent thread the repeated references to flame and madness. It lacks the cohesion of Crowley's work, the grim biblical authority; the text proceeds as a whirlwind stream-of-consciousness rant straight from the darker, sensual regions of Parsons' mind. "Yea it is even I BABALON and I SHALL BE FREE. Thou fool, be thou also free of sentimentality. Am I thy village queen and thou a sophomore, that thou shouldst have thy nose in my buttocks? . . . It is

I, BABALON ye fools. My time is come, and this my book that my adept prepares is the book of BABALON."

Parsons returned from the desert driven to perform the rituals he had been "given," and Hubbard began seeing visions once more. Hubbard was a master storyteller and a quick thinker, but he had now been performing magic with Parsons for nearly two months. According to Parsons' record of the time, Hubbard was exhausted and often left pale and sweating from his exertions. Finally, a fatigued Hubbard saw a vision calling an end to the working. In a fragment from his writings, Parsons, exhausted and exultant, declared his work a success. He believed that Babalon, in the manner of the Immaculate Conception, was due to be born to a woman somewhere on earth in nine months time. "Babalon is incarnate upon the earth today, awaiting the proper hour for her manifestation," he wrote. "And in that day my work will be accomplished, and I shall be blown away upon the Breath of the father."

Feeling the wide-eyed exhaustion of the desert prophet, Parsons declared that he had to return to reality. He would need to distance himself from the OTO if he was ever to get his affairs in order, explaining to Crowley, "I must put the Lodge in [other] hands; prepare a suitable place and carry on my business to provide the suitable material means [money]."

On March 16, 1946, Parsons wrote a group letter to all the members of the OTO, alluding cryptically to the Babalon working he had just performed, "In the coming months the world approaches one of its greatest crises, and Agape Lodge may well have a basic role in this history. I hope and trust that your own part will help to make this role possible, in the time when the Lodge and the world needs you most." This new aeon, paradoxically, was to start with the end of an era. He announced that he planned to sell the house at 1003 and that anyone living there would have to move out by June 1.

The extraordinary group of scientists and occultists, writers and pulp fiction fans, who had been living at 1003, slowly began to disperse. As a condition of the sale (for which he would receive some $25,000), Parsons arranged to move into the old coach house in the grounds. He handed the running of the now homeless Agape Lodge over to a baffled Roy Leffingwell and asked that he be allowed time to recuperate from his magical ordeal. He planned to get his new explosives business off the ground.

While the Babalon working had been taking place, Parsons had been persuaded to forgo Ad Astra and his old friend Ed Forman and to enter a new business partnership with Hubbard and Betty. The aim was similar to that of Ad Astra: "To pool and accumulate earnings and profits of any nature whatsoever, coming from any source whatsoever, and flowing from the capabilities and craft of each of the partners." Thus, any and all profits from their various works—Hubbard's writing, Parsons' Vulcan Powder Corporation—should go towards what they called "Allied Enterprises."

Parsons' attachment to Hubbard, despite the loss of Betty, had only been intensified by the central role Hubbard played throughout his magical workings. This new venture would cement the ménage with Hubbard and Betty and, Parsons hoped, create some much needed stability in his life. He invested close to his entire savings—$20,970.80, largely gained from the house sale—into the company. Hubbard did the same, although his savings, at $1,183.91, were considerably less. Betty contributed nothing. Parsons' financial role in the OTO had long ago proven his generosity, but this new undertaking also seemed tinged with desperation: Despite Candy's presence he still wanted to win back Betty's affections. As he wrote to Crowley, "I think I have made a great gain, and as Betty and I are the best of friends, there is little loss." He was soon proved entirely wrong.

Hubbard came up with a proposal for the new company. He and Betty would travel to Miami to buy three yachts. Once they found crews, they would sail the yachts back through the Panama Canal and sell them on the West Coast at a much higher price. As told by Hubbard, with his naval background, the plan sounded both eminently practical and glamorously adventurous, and Parsons was easily persuaded by his friend's confidence and by Betty's entreaties. Not everyone bought Hubbard's plan, especially not the OTO members who remained at 1003. They feared that their one-time "Rich Man of the West" risked making himself poor very quickly. Grady McMurtry, now based in San Francisco but keeping an eye on Agape Lodge, warned, "It would seem more of an adventure than a business proposition." Jane Wolfe joined in the chorus of dissatisfaction. "I am wondering," she wrote to Germer, "if Ron is another Smith?"

Even Parsons, blind to all suspicion, might have recognized the danger had he seen the letter Hubbard now wrote to the chief of naval personnel, requesting permission to leave the United States "to visit Central & South America & China" for the purpose of "collecting writing material" under the auspices of Allied Enterprises. Hubbard was preparing for a world cruise, not a business trip. Ignorant of these plans, Parsons waved good-bye to his best friend and his ex-lover as they headed east with over $20,000 of his money in their pockets.

However, once out of the direct beam of Hubbard's charisma, Parsons began to lose confidence in the venture. As the weeks passed, Parsons' explosives work dwindled to a halt—he had no money for supplies. He began to worry. He told friends that he was going to persuade Hubbard and Betty to return to Pasadena immediately so they could dissolve the partnership; he realized it had been a mistake to invest all his money in such a scheme. But when he received a phone call from Hubbard, collect from Miami, his manner changed immediately. Parsons succumbed once more to Hubbard's ebullient persuasiveness, swinging from anger and distrust to acquiescence and an almost childlike respect. Louis Culling, one of the remaining OTO members, was shocked to hear Parsons end the conversation "eating out of Ron's hand," telling Hubbard, "I hope we shall always be partners."

The other members of the household saw the dangers of the situation clearly. Culling wrote to Germer and Crowley, expressing his frustration at Parsons' gullibility, "Ron and Betty have bought a boat for themselves at Miami Florida for about 10,000 dollars and are living the life of Riley, while Bro John is living at Rock Bottom, and I mean *Rock Bottom*. It appears that originally they never secretly intended to bring this boat around to California coast to sell at a profit, as they told Jack, but rather to have a good time on it in the east coast."

Crowley needed no more evidence. In a cable to Germer, he cast his judgment: "Suspect Ron playing confidence trick—Jack Parsons weak fool—obvious victim prowling swindlers." When Parsons heard this assessment, he finally shook himself out of his indecisive stupor. With the last of his money, he bought a plane ticket to Miami.

Hubbard and Betty had been busy. They had bought three sailing yachts, the *Harpoon,* the *Blue Water II,* and the *Diane,* and were only waiting for Hubbard's latest navy disability check to arrive before they set sail. Parsons, meanwhile, was hot on their heels. He checked into a cheap hotel in Miami Beach and began scouring the marinas and yacht clubs for any information about the two or their purchases. It did not take long before he traced the sale of the *Harpoon* to a harbor on County Causeway, but Betty and Hubbard were nowhere to be seen. On July 1, Parsons managed to place a temporary injunction and restraining order on Hubbard and Betty to stop them from leaving the county, selling the yachts, or touching any other assets of Allied Enterprises. Now all he could do was wait for them to appear.

After four days of pacing in his hotel room, Parsons received a phone call from the marina. Hubbard and Betty, presumably having heard of his presence in Miami, had rigged up the double-masted *Harpoon* and sailed out of the harbor with the aid of a crew paid for with Parsons' money. He was powerless to stop them. In his hotel room, he drew a magic circle on the floor and stepped into it. He performed "a full invocation to Bartzabel," a ritual for invoking the spirit of Mars to help him in his plight. The spirit failed to materialize, or did it? "At the same time, as far as I can check," he wrote to Crowley, "his [Hubbard's] ship was struck by a sudden squall off the coast, which ripped off his sails and forced him back to port." Was this yet another side effect of his magic? Alerted by the harbor master, Parsons was waiting for the errant pair when they limped back in.

In court the following week, Allied Enterprises was dissolved. The court settlement ordered Hubbard to give Parsons a promissory note for $2,900, and Parsons agreed not to press any further charges—partly, it seems, because Betty had threatened to press charges against him over their past relationship, which began when she was under the legal age of consent. The episode left Parsons shattered. He flew back west. One month later, Betty and Hubbard were married.

When Parsons arrived back in Pasadena, 1003 looked more overgrown and dilapidated than ever. The last of the OTO lodgers had moved out, but not before stripping the gold leaf from the library ceiling. The remaining members of Agape Lodge now met in Los Angeles. From the coach house where he now lived with Candy, Parsons wrote to Crowley and once more offered his formal resignation from the OTO. This time Crowley accepted it.

Parsons completed the sale of 1003 and watched as the new owners demolished it with tractors and wrecking ball. For all his talk of new aeons, of new ways of living, of a whole new morality, the demolition did not feel like a new beginning. The Parsonage, as it had come to be called, had briefly been an adult playground saturated with philosophical hopes and pungent romanticism, fruit brandies and fencing, bohemians and scientists, poetry and rockets. Now it was gone forever. A few months later one of the former residents came to visit the ruins. A copy of

The Book of the Law was found amidst the rubble. It seemed like a tombstone.

Parsons' pursuit of Hubbard had been closely followed by Hubbard's fellow science fiction writers. For L. Sprague de Camp, a Caltech graduate in aeronautical engineering and now one of the most popular science fiction and fantasy writers of the day, the events confirmed his already low opinion of Hubbard. In a letter to Isaac Asimov, he wrote:

> The more complete story of Hubbard is that he is now in Fla. living on his yacht with a man-eating tigress named Betty-alias-Sarah, another of the same kind ... He will probably soon thereafter arrive in these parts with Betty-Sarah, broke, working the poor-wounded-veteran racket for all its worth, and looking for another easy mark. Don't say you haven't been warned. Bob [Robert Heinlein] thinks Ron went to pieces morally as a result of the war. I think that's fertilizer, that he always was that way, but when he wanted to conciliate or get something from somebody he could put on a good charm act. What the war did was to wear him down to where he no longer bothers with the act.

Yet Hubbard was not as washed-up as de Camp had thought. Indeed, Hubbard was about to begin the greatest work of his life. Succeeding exactly where Crowley had failed, he would found a worldwide religion.

For the May 1950 edition of *Astounding Science Fiction*, Hubbard wrote an article entitled "Dianetics—The Evolution of Science." The magazine's editor, John W. Campbell, prefaced the story with a glowing testimonial that praised Dianetics as a truly "scientific method" of mental therapy. Hubbard described Dianetics as a form of psychotherapy that he had discovered through his "dabbles" in mysticism and through a lifetime of mingling with "the shamans of North Borneo, Sioux medicine men," and, most notably, "the cults of Los Angeles." Dianetics depicted the human brain as an "optimum computing machine" in which there are "aberrative circuits"—traumas from the past—introduced to it from the outside world. If these circuits could be swept away, the "optimum brain" could be revealed and the subject would become "clear."

Astounding Science Fiction didn't class Hubbard's essay as fiction, but its language was clearly tailored to the science fiction fan. Like the Charles Atlas bodybuilding advertisements that also ran in the pulp's pages, Dianetics promised to transform the reader's 'normal' brain into an "optimum" brain and thus help man "continue his process of evolution toward a higher organism." Stutters could be eliminated, bad eyesight corrected, learning disabilities overcome, intelligence boosted. Even schizophrenia and criminal behavior could be cured. Although some of Hubbard's fellow science fiction authors might have disapproved of him, Dianetics became an immediate success with the fans. Forrest Ackerman, the heart and soul of the LASFS remembered, "Here in Los Angeles we felt we were going to have a brave new world. That everyone was going to be a 'clear,' were going to take off all their glasses, there would be no more colds, one fella even had a finger missing from a hand and he felt like a chameleon that he was going to be able to grow a new finger."

Within months Hubbard published a book, an expanded version of his essay, which became a national best-seller. It was easy to see why. Dianetics denied the complexities of psychiatry and instead proposed a much simpler model of the mind. Adherents needed no formal education and could begin practicing the techniques after only a few hours of training. Over the ensuing months, thousands converted to Hubbard's creed. The public were invited to come to the newly formed Dianetics Institute and undergo a ten-day "auditing" for only $600. During this period an "auditor" (the therapist) would encourage the patient to relive traumatic shocks (known as "engrams") that they had received as children or even in the womb. Hubbard claimed these shocks were the cause of all mental aberrations in later life. When pinpointed, he explained, the shocks could be erased. The *Los Angeles Daily News* reported that "Hubbard has become in a few swift months, a personality of national celebrity, and the proprietor of the fastest growing movement in the US." Within two years Hubbard had elaborated a religion around his book. Dianetics would become the central text of Scientology, which echoed the themes of the pulps in concerning itself not only with this life but with past lives spent on other planets.

It is hard to ignore certain similarities between Crowley's Thelema and Hubbard's Scientology. Both religions have as leaders charismatic men with logorrheic tendencies. Both preach that man is an immortal spiritual being, that his capabilities are unlimited, and that his spiritual salvation depends upon his attainment of a "brotherhood with the universe." While Thelema was born of the Old World, however, Scientology was distinctly a product of the New. The OTO arose out of the Victorian fascination with mysticism, magic, and the secret societies of Europe. Scientology was a direct product of the twentieth century's childlike trust in scientific knowledge, the success of scientific fantasy, and the Californian desire for self-improvement. Perhaps the biggest difference between the two was in popularity. While Crowley struggled throughout his life to popularize the OTO, the Church of Scientology became hugely successful, and now claims over eight million members in some 3,000 churches spread across fifty-four countries. It is said to make more than $300 million a year, and Hubbard's numerous writings are central to its success. It is, in short, everything Crowley had wanted the OTO to be.

The story of L. Ron Hubbard's involvement with Parsons has proven controversial. After *The Sunday Times* (London) published an article in December 1969 revealing the Hubbard-Parsons connection, the Church of Scientology issued a statement asserting that Hubbard had actually been sent by the United States Navy to live at 1003. He was, it said, under orders to break up the "black magic" cult that resided there. In the process, he "rescued a girl" and "dispersed and destroyed" the group. There is no doubt that Hubbard's arrival at the house on Orange Grove signaled a turning point in the fortunes of both Parsons and the OTO, but whether he acted at the behest of a government agency or because of personal motives is a question best left for the reader to decide.

The summer of 1946 was a ghostly one. The United States Army exploded two atomic bombs off Bikini atoll in the Pacific, one above and one below the water, in order to see what impact they would have on ships. The Nuremberg trials of the captured Nazi leaders were drawing to a close; eleven men were sentenced to hang. From around the world came reports of peculiar sightings. A "ghost rocket" was seen above Denmark, "flying noiselessly at 1,000ft." Three days later a "bluish ball of light" was spotted over Portugal, and the next day a rocketlike object was seen in the skies over Yugoslavia. Back home little could be seen in the skies. An eye-stinging layer of smoke and fumes cast a "mysterious pall" over Los Angeles, reducing visibility to just two blocks.

In October, thirty-two-year-old Jack Parsons moved out of Pasadena. The move was an expulsion from a rapidly fading dream world. Many of the grand estates of Orange Grove were falling to the wrecker's ball. Architectural expressions of outsize personalities were not in keeping with the new postwar austerity. Sprawling plots of land were subdivided and fenced in. Utilitarian conformity inundated grandiose Millionaire's Row. Garden apartments replaced the Busch Gardens; condominiums were built on the grounds of 1003; industry began to flood into the city; resort hotels became business hotels. Pasadena, once the jewel of the valley, was losing its luster. A glance up at the San Gabriel Mountains to the north revealed startling evidence of the change. Where once the mountains had stood imperiously clear against the blue sky, now they were hazed over with smog and growing increasingly indistinct. Indeed, the pollution was enfolding Pasadena into the sprawling Los Angeles metropolis, slowly erasing the city's sovereign and fanciful past.

In an attempt to recoup some of his losses from the Allied Enterprises debacle, Parsons got a job at North American Aviation in Inglewood, in west Los Angeles. The rocket boom triggered by the Second World War had seen the company embark upon the government-funded Navaho Missile Program. Parsons worked in the laboratory, presumably making rocket fuel once more, and he and Candy moved to nearby Manhattan Beach.

Candy's striking looks, particularly her shock of red hair, attracted constant attention, though it was not always complimentary. George Frey, a friend of Parsons from North American Aviation, remembered that one day Candy had returned home quite disturbed. She had been driving around in Parsons' old convertible Packard when she stopped at some traffic lights. Some children at the side of the road began to shout and scream, "Look at the witch! Look at the witch!" before she hurriedly drove off. Parsons found the incident amusing. He took to calling her his "witch" as a term of affection—he had conjured her up after all—and she had all the characteristics of the mysterious and seductive witch mentioned in Jack Williamson's story "Darker Than You Think."

Parsons slowly began to interest Candy in the tarot and astrology. In turn, she painted him pictures—a portrait of Dr. John Dee and a gruesome picture of the errant Betty, her legs cut off above the knees and bleeding. Despite her increasing enthusiasm for the occult, Candy sometimes found Parsons' deep reliance on magic bemusing. One day, she recalled, a windstorm blew off the sea. "It blew the French doors open, and everything was blowing and I'm yelling . . . And can you imagine? I'm running around trying to close the windows, and Jack goes upstairs with his dagger to stop the wind!"

Candy was different from his other loves. She offered neither Helen's patient support nor Betty's joie de vivre. Strong-
willed and erratic, she had no wish to settle for one man's
affections, and she was not about to adopt the passive role of
Parsons' student-lover. When he suggested that their relationship be an open one, it seems that Candy wholeheartedly agreed.
Her indifference seemed designed to help Parsons toughen up
emotionally. As he wrote in his memoir, "Candy appeared in answer to your call, in order to wean you from wet nursing."

Parsons still socialized with the few of his old friends who
remained in the area. He visited Andrew Haley; home videos of
the time show Parsons, Ed Forman, and Haley playing leapfrog
together, while Candy plays with the Haley children and strikes
fencing poses with a stick. But Parsons' manner had somehow
altered, even towards his old friend Forman. "Candy's arrival
had changed a lot of things," remembered Forman's wife,
Jeanne. "Jack wasn't acting the way he used to." He was withdrawn, and his relationship with Candy seemed distant and
lacking the intensity he had shared with Betty. "Jack was an af-
fectionate guy," remembered George Frey, "but they were not
too responsive [together], a bit British." Bob Cornog, who after
living in 1003 had also moved to Manhattan Beach, remembered, "They did not make differences or affinities public. They
were two very independent people living together." Some
thought that Parsons had never escaped Betty's hold. "Cameron
[Candy], she was a nice person," remembered Alice Cornog,
"but she couldn't compare with the other person that Parsons
was madly in love with." When the couple paid a visit to Robert
Heinlein, however, Parsons gave him one of the poems he had
written about Candy. Entitled "Desire," the poem displays a
sensuousness that belies the couple's outward coolness:

> Now I am a whip coiling across your naked buttocks
> Your flesh writhes under my caress, and your voice
> Is shrill with pain and passion

I am a flame that crawls slowly about you
I have found the soles of your feet, and seek
each nerve center.

Although they had known each other for less than a year,
on October 19, 1946, four days after he was officially divorced
from Helen, Parsons and Candy were married. The ever faithful
Ed Forman stood witness.

With his move to Manhattan Beach and his marriage to Candy,
it seemed as if Parsons had put all thought of his time at 1003
behind him. He began selling his vast library of Crowley's books
and occult texts, and he had little contact with the other members of the OTO, except in informing them of the few assets
Agape Lodge had accrued during his time as its leader. Having
lost its home and its charismatic head, the lodge was nearly de-
funct. "At the moment," wrote Jane Wolfe to Grady McMurtry,
"we are penniless, will-less it seems to me, and generally inac-
tive." Wolfe visited Parsons in Manhattan Beach to try to tempt
him back to the order, but Parsons turned her down. November
passed without event; Parsons failed even to acknowledge the
supposed date of the birth of Babalon, the goddess he had summoned earlier that year.

Parsons now seemed diligently absorbed in his job at North
American Aviation. There were a few diversions. In March
1947, the Los Angeles Police Department asked Parsons to provide expert information on a giant explosion that had occurred
at a downtown electroplating plant. The blast had killed 15
people and injured 151. An "atomic-like explosion of smoke
and flame" had destroyed the factory and leveled nearby houses.
Along with eight other experts in chemistry, explosives, and met-
allurgy, including some from Caltech, Parsons investigated the
causes of the explosion. Almost ten years had passed since his
first flirtation with fame at the Kynette trial, but at thirty-two
years of age, he was still the youngest expert on the coroner's

jury. It did not take long for them to trace the eruption to "an exploding cauldron of dangerous chemicals mixed by a young chemist in presumptive disregard of the warnings of science." The young chemist had "masqueraded as a man possessing several scientific degrees, while, in truth, he never finished high school." Parsons surely sympathized with the culprit.

Two months later Parsons gave "a general talk on rocketry" to the Pacific Rocket Society, one of the new breed of amateur rocket clubs started in the wake of the Second World War. These new societies were generally formed not with the purpose of making scientific breakthroughs, but more as a hobby, for the fun of building rockets. In his speech, Parsons predicted that nuclear-powered rockets might enable man to reach the moon; his dreams of interplanetary travel were still keen.

Candy was making plans to travel to Paris in the autumn, ostensibly to study art. Parsons made a talisman for her to wear to protect her from physical and spiritual harm. It consisted of a six-inch piece of rope with nine knots tied in it, a small piece of dark blue felt cut in a circle, two pieces of metal, one seed, a small piece of quartz, a deep-blue stone, and a phonograph needle. He also suggested that she take time to visit Crowley in England. Crowley's wisdom would feed her increasing interest in the occult, and Parsons hoped she might even convince Crowley to forgive his former protégé for his past weakness.

In anticipation of this visit, Parsons wrote a letter to Crowley for the first time since his resignation from the OTO. "It has been almost a year since I last wrote—at that time I was near mental and financial collapse. Since that time I have gained some sort of mental equilibrium and gradually regained something of a position working in my old field in a large aircraft company...My aim is to rebuild myself."

It is impossible to know whether Crowley would have cared for Parsons' tentative attempts to appease him. He never received the letter, nor did he meet Candy. Throughout 1947 his health had been declining rapidly. He was by now injecting himself with as many as eleven grains of heroin a day (the usual dose for most addicts being one-eighth of a grain.) With heroin added to his various other medicines, he was permanently dazed, his once remarkable mind dimmed. Before he had become too weak to act, he had placed Grady McMurtry in charge of the OTO in the United States, under Karl Germer's review. On December 1, 1947, the Great Beast died, aged seventy-two. Some said that he died with tears on his cheeks, his last words being the uncharacteristically hesitant, "I am perplexed." Others thought that his final utterance was the melancholic admission, "Sometimes I hate myself." His obituary in *Time* magazine stated, "The world of 1947 buried him almost without noticing it, and without a shudder."

12

....

INTO THE ABYSS

I remember
When I was a star
In the night
A moving, burning ember
Amid the bright
Clouds of star fire
Going deathward
To the womb.

—JOHN WHITESIDE PARSONS, untitled

The United States had emerged from the war years in better shape than any other combatant country. Since Pearl Harbor the average weekly earnings of the nation's workers had almost doubled—from $24.20 to $44.39; rationing was a distant memory and birth rates were soaring. Nevertheless, the postwar euphoria quickly dissipated as the country found itself embroiled in yet another international crisis. By 1948 Americans had come to view the Soviet Union not as the brave ally of wartime propaganda but as a serious global threat. As the United States mulled over the Marshall Plan, its vast postwar aid package to Europe, the Soviet Union's foreign policy, under the paranoid direction of Joseph Stalin, became increasingly aggressive. In February the USSR invaded Czechoslovakia, and in June the Soviets imposed a blockade upon Berlin, prompting the West's year-long airlift of supplies for the starving inhabitants.

The United States responded to increasing Soviet hostility with extreme measures of its own. The House Committee on Un-American Activities (HUAC), a politically rapacious body charged with seeking out "Red" influence in the country, had been formed back in 1938, but its activities had been curtailed during the war years. Now it was reactivated, beginning with a much publicized investigation into Communism in Hollywood, which had long been charged with placing subversive messages in its films. With the intense pressuring it put on witnesses to name former associates, the vague and sweeping accusations it made against individuals, and its readiness to assume guilt through association, the committee was a formidable terror. Witnesses who refused to answer were cited for contempt of Congress and jailed. Its highest-profile victims by 1948 had been a group of screenwriters and directors known as "The Hollywood Ten," who had refused to give evidence. When the committee accused former State Department official Alger Hiss of being an active member of the Communist Party and of handing government secrets over to the Soviet Union, fear and suspicion spread wider. If Reds had infiltrated the government, where else had they gotten to? Everyone was a suspect. Forrest Ackerman later discovered that the FBI had even planted an informant in the ranks of the LASFS to seek out science fiction fans with Communist sympathies.

Within Parsons' immediate circle of friends, the increasing scrutiny into a person's political history had been causing some concern. In 1946 he had been invited to Frank Malina's for a farewell dinner. Malina had decided to leave America and take a post in Paris with UNESCO. He would be going alone. His consuming dedication to rocketry work during the war had placed his young marriage under an intolerable strain. Now he and his wife Liljan were living apart. Malina had other problems as well. His house had recently been ransacked, although nothing had been stolen; he suspected that the FBI was harassing

him. In New York, Liljan's address book had disappeared from
her car, only to mysteriously reappear a few days later.

The following year Parsons' friend Robert Cornog lost his
security clearance while he was working as a research engineer
at Northrup Aircraft. Without a security clearance it was nearly
impossible to work in an industry dealing largely with classified
government contracts. The rocketeers had all received clearance
for their JATO work, but in the increasingly unpredictable political climate, these lifelines to the world of cutting edge research might any day be revoked. A scientific blacklist had been
created and it took very little to get one's name on it. A disgruntled former colleague might claim one held "liberal social views"
or, as in Cornog's case, might mention one's ties to alleged Communists, and a full-blown FBI investigation would follow. No
explanation was necessary for the withdrawal of a security
clearance. In this Kafkaesque climate, one was accused, tried,
and found guilty without knowing the nature of the crime or
even that one was under investigation. Cornog was released
from his job without knowing why, and, reeling with shock, he
left Los Angeles for Berkeley. It would be almost two years
before he found work again, as an assistant professor of mechanical engineering at the University of California. The job was
theoretical enough not to require a security clearance.

Parsons was naturally worried. He had attended meetings
of a Communist group before the war, and some of his best
friends had been Communist Party members. Fearing harass-
ment or worse from an increasingly Red-scared government,
Parsons began to correspond again with the Suicide Squad's old
mentor, Theodore von Kármán, about the possibility of working abroad.

Kármán was rarely in Pasadena now. As the founder of the
Advisory Group for Aeronautical Research and Development,
he traveled incessantly between the United States and Europe,
attempting to promote international scientific cooperation on
rocketry, among other topics. He told Parsons he would keep

an eye out for any rocketry jobs that suited him, but he also advised Parsons that if he wished to work in what was now an established profession, he needed some academic qualifications, especially in advanced mathematics. Parsons grudgingly enrolled in a night course at the University of Southern California. Going back to school must have seemed strange to him when he had achieved so much without any degrees. At any rate his efforts were rendered futile when, a couple of weeks later, his security clearance was stripped from him with no warning.

Without Parsons' knowledge, the FBI had been investigating his past for months, presumably ever since he had started work on the army-sponsored Navaho Missile Program at North American Aviation. Parsons' clearance had not been reviewed since his work at GALCIT during the war. Now the old Pasadena police reports about the goings-on at 1003 had been unearthed, and someone had charged that Parsons had ties with "an alleged Communist Party member." He was immediately listed as an "Undesirable Employee for National Defense Work."

North American Aviation suspended Parsons from his job immediately. He wrote a panicked letter to Kármán pleading his case: "As you know, I am not a communist, and have no connection with communists or communist front organizations. I have no idea of the reason for this action. Possibly it is simply because I am not enough of a rubber stamp personality . . . Under these conditions, I feel that [it] is desirable to leave this country, and to begin a career elsewhere: in a more liberal atmosphere, as soon as possible. I will greatly appreciate any advice or suggestions that you can offer in the matter. I am really anxious to make the change as soon as possible."

To make matters worse, Candy left him a month after he had lost his clearance. Since she had returned from Europe, their relationship had grown increasingly tempestuous. She had gotten bored playing the role of scientist's wife in Manhattan Beach, and she now planned to go to Mexico to join the artists' colony in the town of San Miguel de Allende. Living there was cheap and the location was beautiful, its sixteenth century houses encircled by mountains. Many Americans, supported by the GI Bill, were now traveling there. San Miguel was renowned for its fiesta atmosphere, its heavy drinking, its bullfights, and its peyote. Parsons was now left alone—without job, friends, or wife.

He began pumping gas at a filling station on the weekends and working as a mechanic fixing cars. He also worked as an assistant in a medical hospital, and he even briefly held a staff position in the Department of Pharmacology at the University of Southern California. He may have won the job because of the expertise he had acquired manufacturing narcotics at home. His closest friend Ed Forman had also fallen into similar difficulties. "After Aerojet he had a really hard time doing anything for a while," remembered his stepdaughter Jeanne Ottinger. "I can remember him sitting up on the roof just flying kites." Unlike Cornog, both men lacked the academic qualifications to fall back on serious theoretical work. Parsons became even less likely to attain them when he failed out of his mathematics course at the University of Southern California. Once again he turned to his magic. If he could not control the real world, then at least he could assert himself in his imaginative magical one.

He embarked on a series of magick rituals, hiring prostitutes or entering into passing affairs in order to carry out his sex magick workings. His new magical endeavor was called "The Crossing of the Abyss," and its aim was to transform him into a "Master of the Temple." At that point his consciousness would supposedly become one with the universal consciousness. The last of Crowley's disciples to attempt the operation was a Canadian accountant named Charles Stansfield Jones in 1916. He had declared himself "Master of the Temple" shortly before being arrested for walking naked through the Vancouver streets.

Parsons carried out his rituals over some forty days of what he described later as "madness and horror." Thoughts of death and suicide possessed him. When he finished, he began frantically writing. Twelve years earlier he had collaborated with

Frank Malina and Edward Forman on a scientific paper called "Analysis of the Rocket Motor." Now he was writing *Analysis by a Master of the Temple,* a histrionic autobiography that recreated the story of his entire life so that this one moment of magical achievement was its climax. Becoming a Master of the Temple allowed Parsons to recast all his disappointments and failures as successes. His parent's divorce, his isolation as a child, his interest in chemistry, the loss of the family fortune, and his betrayal by Betty all appeared as predestined steps on his path to magical fulfillment.

He also began writing a political tract on "liberalism and liberal principles" which he titled "Freedom Is a Two-Edged Sword." In it, he responds to his treatment at the hands of the government by denouncing the increasingly intolerant nature of postwar American society. However, Parsons' remedy to the problem is not social or political reform but the arrival of Babalon, "girt with the sword of freedom." He sought out Wilfred Smith, his former mentor, and declared that he, Parsons, was the Antichrist, his mission "[that] the way for the coming of BABALON be made open."

It seems symptomatic of some form of psychosis that as Parsons' emotional life had been thrown into chaos and his professional achievements retreated further into the past, he should cling to his magic as if it were a raft on a raging sea. His writings of this period contain oblique expressions of deep self-loathing as well as repeated references to all-consuming flames and his own death. Parsons seemed to be casting himself in the role of doomed hero in a cosmic drama that was coherent only to his own mind. Alienated from the OTO, separated from his wife and friends, he seemed to be preaching to himself, declaiming to an empty room, playing to the void.

This period of mania seemed to have eventually worn off and, as 1949 arrived, Parsons seemed to settle himself. He set out, newly determined to regain his security clearance, and consulted with his old friend Andy Haley, who agreed to appeal his suspension. Had Parsons' freshly written "Manifesto of the Antichrist" been discovered, Haley's efforts would likely have come to naught, but the document remained secret. After a closed court session in which Parsons denied any Communist affiliation and defended his involvement with the OTO as a nonpolitical religious organization, the Industrial Employment Review Board reversed the restriction on Parsons' security ranking, claiming the previous judgment had been made "without sufficient cause." He was granted back pay and fully restored to work on classified and top secret projects. Parsons was now back in the real world and functioning with remarkable lucidity.

Kármán had also made efforts on Parsons' behalf. The Hungarian professor had close ties with the American Technion Society, an increasingly powerful organization providing American technological knowledge to the fledgling state of Israel. Kármán had been involved with the society since its foundation in 1945. With his aeronautics connections, not to mention his links to the commanding general of the United States Air Force, he had played an instrumental role in Israel's creation of an air force. Now he put Parsons in touch with Herbert T. Rosenfeld, president of the Southern California chapter of the society.

Rosenfeld was a powerful figure who routinely brokered multimillion dollar contributions from eager donors in the United States to the Israeli government. He met with Parsons and explained Israel's desire for its own rocket program. He asked Parsons to write a proposal for an explosives plant in Israel, and he requested data concerning rockets and other armaments. If Parsons proved he was competent by providing these reports, Rosenfeld would help him leave the United States, as he wanted, and he could open up a whole new chapter in his life, building rockets in Israel.

Since regaining his security clearance, Parsons had left North American Aviation and moved to the Hughes Aircraft Company in Culver City, the company owned by the increasingly reclusive millionaire Howard Hughes, where he worked on chemical plant design and construction. The job put him in an ideal position to gain data for his Technion proposal. By midsummer Parsons had handed over "several reports" to Rosenfeld, all of which were forwarded to Israel for approval. It seems unlikely that Parsons provided any classified material to Rosenfeld. Why, after all, would a man who had just fought so hard to regain his security clearance risk losing it again? Regardless, it was an inauspicious time for an American scientist to collaborate with a foreign power.

The writer L. Sprague de Camp had followed the Hubbard/Parsons drama from afar. Now visiting California, he wanted to meet the scientist-magician he had heard so much about. He wrote to Robert Heinlein for help in arranging an introduction. At the time, Heinlein was working as technical advisor on the film of his story "Rocketship Galileo," now retitled *Destination Moon*, Hollywood's first attempt to give the public a realistic glimpse of the science of rocketry and space travel. "I think he is browned off on the OTO," Heinlein wrote to de Camp about Parsons, "but he may still have some belief in it, i.e. you may find yourself dealing with a convinced cultist." Anyway, he conceded, "Jack is one hell of a nice guy and a number-one rocket engineer."

Parsons fulfilled de Camp's expectations. De Camp was taken by this "big, florid, good-looking, youngish man with the indefinable aura of inherited wealth, who drove an old open Packard with a loose door wired shut." De Camp was researching a book on magic and the occult, and the two spent much time talking together about magic and science fiction. When de Camp asked Parsons about his dealings with Hubbard, Parsons was in good humor enough to admit that he had received a letter from an irate Hubbard, "offering" him Betty back. He then told de Camp that he had summoned his wife, Candy, through magic. When de Camp asked where she was now, Parsons

replied ruefully, "I think she's in Mexico, getting a divorce."
"An authentic mad genius if I ever met one," declared de Camp
afterward.

The admiration of an occasional visitor, however, was not
enough to ease the isolation that plagued Parsons. "He was kind
of lonely at that time," remembered his friend George Frey; "he
got tired of living alone." Down in San Miguel de Allende,
Candy had cultivated a coterie of lovers, both men and women,
including a local nobleman and a bullfighter. Parsons saw nothing of her apart from the occasional fleeting visit. To alleviate
his loneliness Parsons acquired a semipermanent girlfriend, an
Irish girl named Gladis Gohan. He decided to move into a new
home with her, a home that might remind him of his childhood.

The home of 1200 Esplanade stood on the seashore on Re-
dondo Beach. Complete with crenelations, Moorish arches, and
windows framed with stained glass, the house resembled a
Gothic castle. It was, however, made completely out of concrete.
It was as if Parsons' love of epic grandeur and magniloquent
gesture had been distorted in a fun house mirror. Those who visited called it the "Concrete Castle."

When Candy made one of her rare visits to her husband and
found his girlfriend installed in the new house, she paid no heed
to her. George Frey, slightly concerned, asked her what she
thought of the arrangement. "I think she gives the place a nice
feminine touch," deadpanned Candy. Parsons' joy at seeing
Candy again swiftly dissipated as the two began arguing fero-
ciously. Candy finally left for Mexico once more, and an embit-
tered Parsons initiated divorce proceedings against her on the
grounds of "extreme cruelty."

A rocketeer reunion took place in Pasadena on June 6, 1949.
Frank Malina had returned to the United States from France to
visit his old friends and family. He brought with him his new
wife, Marjorie, whom he had met while working at UNESCO.

Andy Haley threw a lavish party on Orange Grove, where the drink and arguments were as copious as ever. In the midst of the celebrations, Haley coaxed Parsons up onto the little balcony in front of his bedroom window, as he had done so many times before. He demanded to hear the "Hymn to Pan" once more. Parsons took his position and began to declaim. This time there were no thrown bottles, no jeering. Everybody stood silent as Parsons recited the verses he had recited so many times.

> The great beasts come, Io Pan! I am borne
> To death on the horn
> Of the Unicorn.
> I am Pan! Io Pan! Io Pan Pan! Pan!
> I am thy mate, I am thy man,
> Goat of thy flock, I am gold, I am god,
> Flesh to thy bone, flower to thy rod.
> With hoofs of steel I race on the rocks
> Through solstice stubborn to equinox.
> And I rave, and I rape and I rip and I rend
> Everlasting, world without end,
> Mannikin, maiden, Maenad, man,
> In the might of Pan.
> Io Pan! Io Pan Pan! Pan! Io Pan!

Parsons finished the final line and a silence fell over the group. They no longer laughed at "Jack being peculiar." Instead they recalled more glorious, more innocent times, a world where the moon was the limit. "I shall never forget Jack doing this," Malina recalled many years later. It would be the last time he saw his old friend.

Malina left America early on June 15. He did so just in time. That same day the *Los Angeles Times*' front page led with the headline, FRANK OPPENHEIMER ADMITS HE WAS RED. With Sidney Weinbaum, Oppeheimer had presided over the Communist salons that the rocketeers attended before the war. Now the newspapers told how Oppenheimer had "joined the Communist Party while he was working on his PhD degree at the California Institute of Technology at Pasadena." A few days later came the announcement that, along with an accusation against Weinbaum, "the Committee had information indicating that Frank Malina, identified as a former secretary of the Aerojet Engineering Corp. at Pasadena, was a Communist Party member and that Communist 'cell' meetings were held at the Oppenheimers' and at Malina's home."

In early 1950 Klaus Fuchs, a German-born physicist who had worked on the development of the atomic bomb in Britain and the United States pleaded guilty to charges of passing scientific secrets to the Soviet Union. He was imprisoned for fourteen years. Soon Julius and Ethel Rosenberg would be arrested for passing secrets from the Manhattan Project to the Soviet Union. Scientists, the heroes of the last war, now appeared as enemies of the state. Parsons looked on in despair. "Science, that was going to save the world back in H.G. Wells' time, is regimented, strait-jacketed, scared shitless, its universal language diminished to one word, security," he lamented.

He must himself have been scared. Rosenfeld had fallen ill and Parsons' negotiations with him about a move to Israel had "broken down." Since regaining his security clearance, Parsons had been interviewed by the FBI on several occasions about his links to Weinbaum's Communist group. Parsons had initially prevaricated, saying that the meetings he had been to in 1938 and 1939 "were in fact study groups and dealt with many 'isms' other than Communism." Now, however, he told the FBI that Weinbaum held "extreme communist views" and "knew of the existence of a communist group on the Caltech campus." In a climate of rumor, such words were proof and clinched the FBI's case against Weinbaum. On June 16 Sidney Weinbaum was arrested for perjury. He had signed a Caltech security form stating that he had never been a member of the Communist Party.

The reasons for Parsons' betrayal of Weinbaum are not entirely clear. The threat of losing his security clearance again must have weighed heavily upon him, considering the depression his lost job had provoked before. Perhaps he thought Frank Oppenheimer's confession had already doomed Weinbaum. Or perhaps Parsons still harbored a grudge against Weinbaum for breaking up the discussion group he and Malina had founded all those years ago.

In any case, Parsons was not the only member of the Suicide Squad to be troubled by the FBI. Martin Summerfield lost his security clearance for his affiliations with known Communists, and even the respected Kármán came under investigation for his links to Béla Kun's Communist regime in Hungary in 1919. Tsien, however, became the most spectacular casualty. He had recently returned to Pasadena to become the first Robert Goddard Professor of Jet Propulsion at Caltech. Following a grilling by the FBI about his alleged Communist sympathies, his security clearance was revoked almost immediately. At a time when 90 percent of all research at the Jet Propulsion Laboratory was classified, he could no longer hold his prestigious position. Tsien made plans to travel back to his Chinese homeland, but when the government discovered his decision, it moved swiftly to stop him. Customs officials searched his packed luggage and discovered papers marked "Secret" and "Confidential." Many of these documents were written by Tsien himself, and it must have seemed natural to him to take them. But the day that Sidney Weinbaum was sentenced to four years without parole for perjury and fraud, "Caltech's famed Dr. Tsien" was arrested and held without bail. He was forbidden from leaving the country for five years so that his scientific knowledge would become obsolete and therefore useless to China's new Communist regime.

As his old friends and acquaintances rapidly fell around him, Parsons contacted a recovering Herbert Rosenfeld and renewed his efforts to move to Israel. Rosenfeld greeted him warmly and asked for one last piece of information. If he provided it, Parsons could leave for Israel immediately.

This final proposal was very simple indeed: Rosenfeld wanted a detailed breakdown of equipment costs for the jet propulsion development program and explosives plant proposal that Parsons had formulated the previous year. Such costs were easily calculated by consulting similar proposals Parsons had been working on for Hughes. He took a bundle of handwritten documents concerning explosives and rocket propellant manufacture from his office and gave them to Blanche Boyer, a female colleague from Hughes, asking if she could type up copies of the material for him. Confident in his plans for the future, he had even included within the bundle "tentative points" for a two-year employment contract he was planning to give to Rosenfeld.

Unfortunately for Parsons, in light of the increasingly hysterical press coverage of "Red" scientists, it seems that the sight of these documents sent Boyer into a panic. No sooner had Parsons gone than she alerted the security authorities at Hughes to a case of possible espionage. On September 26, Parsons was fired from his job at Hughes for removing confidential papers without permission, and the FBI descended upon him yet again. He told them everything. He told them about the Technion Society, he told them about Rosenfeld, he told them about the papers he had already submitted to Israel. He claimed that he was only preparing a cost breakdown from the papers he had taken from Hughes. He swore that he was going to submit the proposal to the State Department and insisted that he wanted only to transfer the costs of equipment to Israel, not the classified material. He pleaded that his action was an oversight, a slipup.

His carelessness caused his past life to be exhumed once again. As the FBI prepared to prosecute Parsons for espionage, they interviewed old friends and enemies, seeking evidence that he was a spy. More stories about his antics at 1003 crept out of the woodwork. An unnamed source claimed that at a weekend party he was drugged and initiated into the order against his will. Others described Parsons as a "crack-pot" and "a religious fanatic." One source focused on his relationship with Candy, describing them as "an odd and unusual pair in that they do not live by the commonly accepted code of married life and are both very fascinated by anything unusual or morbid such as voodoo-ism, cults, homosexuality, and religious practices that are 'dif-ferent'." Some of Parsons' scientist friends rallied around him, explaining that during his time at Caltech security regulations had been very lax and "it was common practice to remove re-ports for [one's] own purposes." But the evidence against him was overwhelming. Making matters even worse, Herbert Rosen-feld himself was under investigation by the FBI for his links to known Soviet agents. When Parsons was pointedly asked about his own political affiliations, he furiously asserted that he was not a Communist but "an individualist." With the wealth of in-formation collected against him, his declaration was futile.

Parsons could do little but wait for the investigation to be over. He became increasingly disturbed. He talked to George Frey about the FBI, referring to them as the "Black Brother-hood." He became convinced that the man living in the next door apartment was spying on him. "Just be careful what you're saying," Frey remembered him whispering; "the apartment is bugged." The one thing he was certain of was that his chance for escape to Israel was gone.

Down in Mexico, Candy heard the news of his FBI troubles. Per-haps she was sympathetic to them, as she now returned to Los Angeles for good. The pair abandoned their divorce proceed-ings and moved from the Concrete Castle to a real place of solace, Orange Grove Avenue. The coach house of the old Cruik-shank estate at 1071 South Orange Grove was for rent. Though the estate itself had been demolished years earlier to make way for condominiums, down the serpentine driveway the coach house was in a world of its own. Across the street the couple could see the brick-and-stone French manor house owned by

John S. Cravens, his third on the avenue, and directly opposite the coach house was the grand Macris estate. Jack and
Candy's new home even stood on the same block 1003 had once
graced.

The ground floor of the coach house, formerly the stables,
had been converted into a huge laundry room. Parsons made it
his home laboratory, storing his chemicals and powder up
against the wall, brewing absinthe next to his explosives. As the
FBI investigation rumbled on and Parsons awaited the possibility of an espionage charge against him, he went back to the employment of his youth, working at the powder companies. By
May 1951, he had even set up his own—the Parsons Chemical
Manufacturing Company in North Hollywood. It was not long
before the couple began to throw parties at the house. Guests
included a wealth of bizarre transients—a former British secret
service agent, a circus performer from Europe—as well as old
friends like Robert Cornog and Ed Forman. It was said that the
jazz legend Charlie Parker came to one of the gatherings with
his girlfriend. Candy's painting of Parsons as the "Angel of
Death," some six feet tall and four feet wide, watched ominously over proceedings. Parsons began to regain something of
his old reputation on Orange Grove. Sometimes the group
would migrate to the pergola outside, and play bongos "until 6
the next morning," remembered some of the guests, "and the
police cars kept coming."

One guest was Greg Ganci, an actor and artist in his
midtwenties, who had met Parsons and Cameron through mutual friends. He had spent some time in San Miguel de Allende,
but after fourteen months of fiesta he had returned to Los Angeles and was now acting at the Pasadena Playhouse along with
his partner Martin Foshaug. "We had a friendly relationship
with Jack, we talked a lot," he remembered, "and Jack was always talking about Gilgamesh and the Babylonian gods and
goddesses and things like that and I would just listen to it with a
tin ear." They did share an interest in Aleister Crowley, whose

reputation had been somewhat revived by the recent release of John Symond's scandalous biography, *The Great Beast*. The tales of Crowley's debauchery and bohemian lifestyle appealed to the new postwar Beat generation, who went out of their way to reject the social conformity of the previous generation, and Parsons entertained his guests with tales of his own about his former mentor.

Despite the sociable atmosphere at these parties, Parsons remained somewhat aloof. He was by now part of an older generation that had thrived on classical music rather than jazz, on piano duets instead of bongo playing. He had become a prewar relic of sorts. Ganci remembered, "[There was] nothing artistic about Jack, he could have passed as an executive of a company. And for us in our twenties—Jack was 35, 36—we considered him much older. He was never really bohemian, everybody else was bohemian." Little did Ganci know the full extent of Parsons' past debaucheries.

If Parsons was still indulging in drugs, he was doing so discreetly. He never showed any sign of undue intoxication to his friends, but down in his laboratory he kept a number of hypodermic needles. It is unclear whether they were for his chemical experiments or, like Crowley before him, for his own private use. He was certainly emulating Crowley in his bulk. He was growing increasingly heavyset, as if the wired energy of his youth had deserted him. But he was keeping busy. He had been formulating a new religion, one that would replace the "claptrap" and "indirection" that he felt had compromised the OTO. His religion was to be created for the "modern spirit" and consist of "an austere simplicity of approach." He named it the "Witchcraft"; it seemed to combine elements of Crowley's teaching with Parsons' own Babalon prophecy, while also drawing heavily on Parsons' favorite Jack Williamson story, "Darker Than You Think." He priced a basic course of instruction at ten dollars; a modest sum when compared to the six hundred dollar cost of a ten-day Dianetics program.

Parsons wrote to de Camp happily, "The witch [Candy] is back with me, quiescent but unexstinct [sic], and I am in the comparative safety and more than comparative respectability of the explosives business." This new found stability promised to endure. On October 25 the assistant United States attorney at Los Angeles declined to prosecute Parsons "due to lack of sufficient evidence of intent or reason to believe that information obtained was to be used to injure the US or to the advantage of a foreign nation." They had obviously concluded that he was more of a crackpot than a Communist. The Industrial Employment Review Board was not so kind. In January 1952 they revoked Parsons' security clearance forever and withdrew his access to "Department of Defense classified information and/or material." The reason stated was simple: "That you do not possess the integrity, discretion and responsibility essential to the security of classified military information . . . that you might voluntarily or involuntarily act against the security interest of the United States and constitute a danger to the national security." Though he had escaped prosecution, any hopes Parsons may have harbored of returning to the field of rocketry were ruined.

The news precipitated another bout of magic. His mood alternated between "manic hysteria and depressing melancholy." He seemed to recover swiftly, however, and started considering a new adventure—a long trip to Mexico. He mused to friends that he could grow grapes down there to make brandy with, or perhaps even build a pyramid to "re-establish the ancient glories." He visited an elderly Jane Wolfe in February 1952 and they talked for three or four hours. He admitted he was afraid of the future but felt he had little choice. "I am finished here; there is nothing more for me to do, so I am going away," he told her, before smiling and saying, "But I shall come back."

He began to prepare himself financially for the trip and contacted Kármán once more, asking him for a reference. Kármán, who had heard of Parsons' misfortunes, was only too willing to help. "I have known Mr. Parsons for more than 15 years," he wrote. "He worked under my direction on rocket problems. His technological knowledge and experience, especially in explosives, is excellent." Nevertheless, the old memories of Parsons at Aerojet could not be completely forgotten. "Mr. Parsons is very inventive; he is better for independent work than for office work with many employees about."

Parsons' temporary job was at the Bermite Powder Company in Saugus, the same town he had worked in as a young man while employed at the Halifax Powder Company. Here he put his remarkable explosives skill to work developing and manufacturing pyrotechnics and explosives for the motion picture industry. There was a boom in war films at the time, and realistic explosives were needed more than ever. He specialized in squibs, small explosive devices placed beneath clothes to simulate the action of a bullet striking the body. But Parsons soon lost interest in the work. The pull of Mexico became stronger. It would be a new start, a chance to clear the mind after the last troubled year. "This was kind of a second honeymoon or something," remembered Ganci, "kind of a reconciliation with Candy. They could go down to Mexico where life was one big fiesta." They planned to leave in mid-June and stay in Mexico for three months, maybe six, just as soon as Parsons had fulfilled his short-term obligations to Bermite.

After all his years working with explosives and rockets, Parsons had accumulated a vast collection of rare and volatile chemicals. Far too unstable to keep in his home laboratory, he had been storing them in a warehouse owned by the Special Effects Corporation, a pyrotechnics company that Parsons often leant his services to, based in Pacoima in the San Fernando Valley. While visiting his stockpile, Parsons ran into an old JPL acquaintance, Charles Bartley. Bartley had founded The Grand Central Rocket Company, a direct competitor with Aerojet, and he had hired some space at the Special Effects plant for his own explosives work—the very space Parsons had been using as a storeroom for his substances. He was instructed he would have to move his supplies. Even Bartley was amazed by the amount and variety of chemicals Parsons had collected over the years, commenting, "Well, he had a lot of explosive stuff that I didn't even know the names of." Parsons' stores included cartons of nitroglycerin, trinitrobenzene, and penthaerythritol tetranitrate, better known as petn, one of the most powerful explosives known to man.

The last time Bartley had seen Parsons was in 1944 in a tin shack on the dried-up riverbed of the Arroyo Seco, when Parsons and Forman had barged into his office bragging about the new Laundromat business they were going to found. "How's that Laundromat business?" Bartley now asked. Parsons turned to him with a gleam in his eye and said that he was now going to start a "fireworks company" in Mexico. With that he began to transport his bottles and beakers, his test tubes and ovens, back to his house on Orange Grove.

With typical casualness, Parsons stacked the boxes in the coach house's laundry room. Since the Mexico trip was almost upon him, he decided to keep many of the bottles in the cardboard boxes he had moved them in. Strictly speaking, it was against the law to keep all these volatile explosives in a residential abode, but he had no place else to put them and who would know about it in the long run?

By the beginning of June, he and Candy had given the lease on the house over to Ganci and Foshaug, although Parsons kept the ground floor for his chemical experiments. The two men had moved in, accompanied by their twenty-one-year-old friend, Jo Ann Price. To celebrate their arrival, they began repainting the house. In honor of their landlord's demonic past, Foshaug painted a large devil with its mouth open above Price's bed. Parsons and Cameron moved in with Parsons' mother, who was house-sitting just a mile or so from Orange Grove at 424 Arroyo Terrace. The new corridor was filled with their bags and suitcases, but Parsons left his "magical" box, covered in runes and Enochian symbols and containing most of his writings, back in the garage of the coach house.

On June 16, the night before Parsons and Cameron were meant to leave, Parsons went for a walk in Exposition Park with George Frey. He excitedly told Frey of their plans to visit the Casa del Inquisidor in San Miguel de Allende, which had a secret tunnel running to a nearby nunnery and was said to be haunted. He predicted that he and Candy would spend several months in Mexico. To Frey he seemed thrilled.

The next day Parsons got a call. It was from the Special Ffects Corporation. They wanted to know if he could prepare a rush order for them before he left. He agreed and went over to the coach house to begin work, leaving Candy to finish the shopping for their trip. Their car, the battered old Packard, now pulled a trailer filled with everything they could possibly need: artist's supplies, paints, canvas, archery equipment, and fencing foils.

By 4:30 P.M. Parsons was still at work. Leaving the house, Jo Ann Price stopped to ask what he was doing. He told her he was mixing some "very expensive chemicals" and was in a bit of a hurry to finish. Most of his chemicals and equipment still lay packed in the boxes. With no beakers and flasks at hand, he was using a tin coffee can as a mixing bowl.

At around 5:00 P.M. Ganci wandered downstairs from his bedroom. Parsons was still working. "He had test tubes and he was pouring liquids from one to another and then putting them into the oven and then closing the oven and then waiting around." After chatting briefly, Ganci turned to go back upstairs. Martin Foshaug and his mother were cooking supper. With a joking look of concern, Ganci turned to Parsons and said, "For God's sake Jack, don't blow us up!" Parsons just chuckled and yelled after Ganci as he walked up the stairs, "Don't worry about it!"

THE MAN IN THE MOON

Along a parabola life like a rocket flies
Mainly in darkness, now and then on a rainbow

—ANDREY VOZNESENSKY, Parabolic Ballad

There was no funeral for Jack Parsons. The OTO held a service to honor the fact that "Brother John Whiteside Parsons has taken his last journey with the sun." Betty, now divorced from Hubbard, was in attendance. Parsons' body was cremated and his widow, Candy, took the ashes into the Mojave Desert. At the intersection of the two massive, whirring power lines, she scattered his ashes and watched as a light desert wind swept them into the air like smoke.

In the years after his death, Parsons' reputation would be distorted by gossip and hearsay. Speculation grew wilder after the discovery of his Babalon writings, with their frequent invocations of his own death and transformation into "living flame." The less people knew Parsons, the more outrageous their claims. Some claimed he had been destroyed for his magical ambitions, and there was even talk that he had been assassinated by an angry Howard Hughes, in retaliation for the former employee's theft of company documents. Some of those who had lived at 1003 with him thought he had never recovered from the pain of Betty's departure, and Jane Wolfe thought his death was a suicide "with the help of the Unconscious." Ed Forman, devastated by the death of his oldest and closest friend, came up with probably the most plausible answer: "Jack used to sweat a lot and the damn thing just slipped out of his hand and blew him up." In life, Parsons had solved his share of mysterious explosions. It was somewhat ironic that his fate should now be inextricably linked to one.

The House Un-American Activities Committee continued its inquisition. Frank Malina was indicted in absentia for failure to disclose his former Communist Party membership on a security form, and his United States passport was revoked. Despite his unstinting efforts during the war to further the cause of American rocketry—efforts which had caused the breakdown of his first marriage—he was effectively exiled by the United States. His strong moral sense was offended as he watched the German V-2 scientists, led by Wernher von Braun—war criminals in his eyes—welcomed by his home country. They would spearhead the postwar United States missile and space program.

Malina remained in Paris, working for UNESCO, and he continued to keep a close eye on the international astronautical scene until good fortune eventually came his way. He had been the only founding member of Aerojet not to sell all his stock in the company back in the 1940s. As the company grew, employing some 25,000 people by the midfifties, Malina unexpectedly became a very rich man indeed. Aerojet eventually became a giant aerospace and defense contractor. Among its many creations was the propulsion system for the Apollo space flights. Malina devoted the rest of his life to painting and became one of the foremost proponents of kinetic art. His work's central theme was outer space.

Tsien Hsue-shen also became an exile. Forbidden to leave the United States, he began to work on the problems of space flight and rocket-powered passenger ships, but he was continually harassed by the authorities and forbidden to work on cutting edge classified material. After five years, the authorities decided to deport him; by that time he was more than happy to go. He arrived in China in 1955 and received a hero's welcome. If he had not been a Communist before, he was soon to become one, and his anger over his treatment at the hands of the United States government persisted. Over the next forty years he supervised the development of China's intercontinental ballistic missile program becoming a national icon in the process. Such was his standing that Chairman Mao himself asked Tsien to tutor him in science. He never returned to the United States and is believed to be alive today. He refuses to speak to western journalists.

In the course of the HUAC purges, Martin Summerfield lost his security clearance because of his familiarity with known Communists, but his brilliance guaranteed him a successful academic career, and he became professor and director of the Combustion Research Laboratory at Princeton University. Apollo Smith, who had never participated in Sidney Weinbaum's Communist meetings, stayed at the Douglas Aircraft Company for most of his life, in the role of chief aerodynamics engineer.

Ed Forman was not so fortunate. The grief he felt following his closest friend's death, combined with a growing bitterness at his exclusion from the increasingly successful Aerojet, changed him. "Ed became a totally different person," remembered his wife Jeanne, as her previously outgoing husband became increasingly withdrawn and aggressive. As boys Parsons and Forman had come to an agreement that should one of them die, that person would try and contact the other from the spirit plane. Two years after Parsons' death Forman went for a night drive in the desert. He would later confide in his family that as he drove he felt a presence in the back seat of the car that he was sure was Parsons. "He knew it was Jack although no words passed between them," recalled his daughter, Lynne. "It really scared him. After that he really didn't want to communicate with Jack again."

Forman's occult books remained on the family bookshelves in amongst copies of his favorite Edgar Rice Burroughs science fiction, but when one of his daughters asked him about magic he replied very seriously, "It's all real, it all works. Don't touch it. You'll get yourself in real trouble." Having lost most of the money he had gained from the Aerojet sale, Forman and his family moved to northern California where he found work as a test engineer on missile systems for Hughes Aircraft and Lockheed. But no matter how far he got from Pasadena the events at 1003 had left an indelible mark on him. His wife recalled how every now and then he would look anxiously around him and ask whether she could hear a long, persistent wailing sound. She never could. The screaming of the banshees he had conjured that night with Parsons stayed with him until his death in 1971 at the age of sixty.

The most mysterious and transient member of the rocket group later made a brief reappearance. Weld Arnold, the enigmatic benefactor who in 1937 gave the fledgling Rocket Research Group $1,000 in one- and five-dollar bills, resurfaced in Nevada in the 1960s. He had left rocketry behind and now presided over the Reno Magic Circle, an organization "open to youths of good moral character interested in the magical arts." Like any good magician, he never revealed the secret of where the money had come from.

After her husband's death, Marjorie "Candy" Cameron grew ever more involved in the occult. She became convinced that she was the incarnation of Babalon that her husband had prophesied, and insisted on burning the majority of his former possessions. She achieved a degree of fame within the underground arts movement in California over the next forty-five years, thanks to her unsettling, powerful paintings, and her pictures appeared in exhibitions across the country. She also performed in a number of avant-garde films, most notoriously acting alongside Anaïs Nin in Kenneth Anger's Crowley-inspired *Inauguration of the Pleasure Dome*.

The hedonistic lifestyle fostered in Parsons' Agape Lodge of the 1940s was revived on a much larger scale by the hippie movement of the 1960s. As for Crowley, his reputation grew and grew. His gospel of "Do what thou wilt"—modified and transformed—appealed strongly to the socially liberated sixties generation. He resurfaced as a countercultural icon; his photograph appeared on the cover of the Beatles' album *Sgt. Pepper's Lonely Hearts Club Band,* and his ideas influenced everyone from Dr. Timothy Leary to the rock group Led Zeppelin. He was hailed as a prophet before his time for bringing together eastern and western esoteric traditions, and although he could never quite escape the "Satanist" tag that he had gained in the Edwardian newspapers, this ensured his present-day popularity.

Wilfred Smith, Parsons' original guide into the occult world, moved to Malibu with Helen and their son, Kwen. From here Smith tried vainly to hold the disparate threads of the old Agape Lodge together, but the frailty of Jane Wolfe and the increasing obstreperousness of Karl Germer meant Smith was doomed to fail. He died of cancer in 1957. Helen would continue to disseminate Crowley's writings through her own publishing company, Thelema Publications. She would remain a self-confessed Thelemite until her death in 2003.

After languishing for years, the OTO was eventually revived through the efforts of Grady McMurtry in the nineteen-seventies. Aided by an upsurge of interest in mysticism and the occult and with the help of Helen Parsons Smith and the few remaining members of Parsons' Agape Lodge, it has since become an international organization with several thousand members worldwide. Since the scientific community had largely overlooked Parsons, the OTO became sole guardians of his story. He became the Che Guevara of occultism, his few surviving writings pored over, his magickal workings the subject of intense debate.

Science fiction, or sci-fi as it became known (the new phrase purportedly coined by Forrest Ackerman himself), swept the world, no longer the diversion of a quixotic (and risible) few.

Parsons was not entirely forgotten by those writers who had known him. Some suggest that much of Robert Heinlein's later work, including his cult classic *Stranger in a Strange Land,* was influenced by Parsons' and Crowley's philosophies. Parsons made only one more appearance in fiction, this time in L. Sprague de Camp's 1956 novelette *A Gun for Dinosaur.* The story recounts the adventures of time-traveling big game hunters who journey 85 million years into the past in order to shoot dinosaurs. The resemblance between Parsons and "Courtney James" is not terribly close—Courtney James professes no scientific ideas or occult beliefs—but the character holds some flourishes of Parsons' manners. Courtney James is a "a big bloke . . . handsome in a florid way, but beginning to run to fat." When asked about his wife, he replies exactly as Parsons had eight years before: "My wife is in Mexico, I think, getting a divorce." And perhaps Courtney James' violent death was modeled after Parsons own. "On the boulevard, just off the curb, lay a human body," de Camp writes. "At least it had been that, but it looked as if every bone in it had been pulverized and every blood vessel burst. The clothes it had been wearing were shredded."

The years following Parsons' death saw a rush into space. In 1957, the Russian satellite *Sputnik* became the first object made by man to orbit the earth. Only then did the general public begin to admire and eventually fear the rocket. Few were cognizant of the genesis of this stunning achievement. Man once more endeavored, as Parsons put it, "to claim his ancient heritage, the stars." The Jet Propulsion Laboratory, fruit of the Suicide Squad's early experiments, played an essential role in the United States space program. The JPL conceived and produced robotic spacecraft to explore other worlds. They manufactured lunar landers and Mars landers, as well as the *Voyager 1* and 2 spacecraft, which have traveled farther from earth than any other human-made objects. Twenty-seven years after their launch, *Voyager 1* and *Voyager 2* continue to follow their paths out of the solar system. In 2003, the JPL employed some 5,500 scientists and had an annual budget of approximately $1.4 billion.

On Halloween in 1968, a memorial plaque was unveiled at the Jet Propulsion Laboratory to commemorate the thirty-second anniversary of the first rocket experiment in the Arroyo. The Laboratory had grown almost exactly on the same spot, though a sprawling complex of car parks and ten-story buildings had long since replaced the corrugated-iron sheds. As part of the celebration, the JPL reconstructed the original rocket testing apparatus on the lawn. Next to the modern towers it looked primordial, like an amoebic monster that had dragged itself out of the baked riverbed of the Arroyo. By this point anti-Communist paranoia had abated. Frank Malina had been reissued a passport and had been allowed back into the country in order to give a speech to the assembled dignitaries and surviving rocketeers. He talked of the group's struggles, of the hostility that had first greeted them, and of the hours spent groping in ignorance trying to establish the basics of a new science. For many in the audience, it must have been difficult to imagine a time when rocketeers had to fight for recognition. "In conclusion, I would like to pay homage to Jack Parsons," declared Malina at the speech's end, "who made key contributions to the development of storable propellants and of long duration solid propellant engines that play such an important role in American and European space technology. He has not received as yet his due for his pioneering work." With the help of the International Astronomical Union, Malina was to make sure that Parsons' work would be commemorated in some small way.

The moon is a crowded place. Of the craters that mark its surface, hundreds have been labeled by the International Astronomical Union. They bear the names of biologists, astronauts, astronomers, and of course, rocket scientists from Russia, the United States, France, and Germany. Albert Einstein and Robert Oppenheimer have craters named after them, as do Theodore von Kármán, Robert Goddard, and Hermann Oberth. On the far side of the moon, the dark side of the moon, between fissures dedicated to Aleksej Krylov, the Soviet mathematician, Paul Ehrlich, the German Nobel laureate for medicine, and Sir John Cockcroft, a physics Nobel laureate, lies a crater forty kilometers wide, named simply "Parsons." It is remarkably fitting that the moon, in many ways the guiding light of Jack Parsons' life, should be the one place where he would finally find acceptance among his peers.

ACKNOWLEDGMENTS

A great many people have helped me in the preparation of this book. I would particularly like to thank the following, without whom it could not have come into being: Forrest Ackerman; Jon Ausbrooks; Jim Bantin at the University of Nevada, Reno; Ray Bradbury; Gene Bundy at Eastern New Mexico University; Russell Castonguay and Charles Miller at the Jet Propulsion Laboratory Archives; Harold Chambers; Dr. Francis Clauser; Robert Clifton; Jesse Cohen; Walter Daugherty; Simon Elliot at UCLA; Shelley Erwin at the Caltech Institute Archives; Cliff Farrington at the University of Texas, Austin; George Frey; Greg Ganci; Ed Green; Alice Greenberg; Hugh Griffin; Christoph Hargreaves-Allen; Carson Hawk; Bill Heidrick; Tony Iannotti; Zach Isaacs; Bo Ketner; Professor Hans Liepmann; Charles Parsons; Delphine Haley; Andrew Haley Jr.; Lynn Maginnis; Marjorie Malina; Dr. Roger Malina; Bob Null at the Los Angeles Science Fantasy Society; Jeanne Ottinger; Lian Partlow at the Pasadena

Museum of History; Bill Patterson; Pepe Rockefeller; Professor William Ryan and Dr. Jill Kraye at the Warburg Institute; Edmund V. Sawyer; Phyllis Seckler; Lloyd Shier; Margaret Sollito at John Muir High School; Tom Sprague; Homer J. Stewart; Dace Taube at the University of Southern California; Karen Thomas at the American Institute of Aeronautics and Astronautics; Guy Walters; Thomas Whitehead at Temple University; Jack Williamson; Lisa Willingham at Texas A&M University; Frank Winter at the National Air and Space Museum; Jan Wunderman; Dr. Benjamin Zibit; Marjorie Zisch.

Special thanks are due to my agent, Jill Grinberg, for her tireless encouragement; my editor at Harcourt, Andrea Schulz, for her much-needed guidance; Dan Fox, who first pointed me in the direction of Parsons; the late Dr. John Bluth, who acted as a guide and proofreader and who was the preeminent authority on the GALCIT Rocket Research Group; the OTO and Thelema Media LLC for allowing me to quote freely from the writings of both Aleister Crowley and John Whiteside Parsons; Scott Hobbs, who selflessly allowed me access to his Marjorie Cameron archive; Martin Starr, who was generous with both his time and scholarly knowledge; Hymenaeus Beta for his constant advice and support; and the late Helen Parsons Smith for sharing her company. Susan Pile was not only a friend and guide, but her painstaking work (in collaboration with Brad Branson) in the creation of her documentary on Parsons and her generosity in sharing it with me, was of untold value.

My greatest thanks are, however, reserved for Charlotte.

SOURCE NOTES

Because of Jack Parson's significant writing disability, all quotations taken from his papers have had their spelling and punctuation corrected.

ADASTRA Adastra LLC, private collection
CALTECH California Institute of Technology archives, Pasadena
DE CAMP De Camp Collection, Harry Ransom Center, University of Texas at Austin
JPL Jet Propulsion Laboratory archives, Pasadena
MALINA Marjorie Malina, private collection
OTO Ordo Templi Orientis archives, New York
STARR Martin Starr, private collection
THELEMA Thelema Media LLC, private collection
YORKE Gerald Yorke Collection, Warburg Institute, London

PROLOGUE

"look at the devastation that surrounds them." Ganci, author interview, 4 August 2003; Greg Ganci and Martin Foshaug, interview by Brad Branson and Susan Pile, 2 September 1995. JPL.

"Let me know the misery." *Pasadena Star-News,* 18 June 1952.

"charges of strange cultism." *Los Angeles Mirror,* 19 June 1952.

"John W. Parsons, handsome 37-year-old." *The Pasadena Independent,* 19 June 1952.

"SLAIN SCIENTIST PRIEST." *The Mirror,* 20 June 1952; *Los Angeles Examiner,* 20 June 1952.

"He worked carefully." *Los Angeles Times,* 21 June 1952.

"summoned a fire demon." Forrest J. Ackerman, author interview, 22 October 2002.

"liked to wander." *Los Angeles Times,* 19 June 1952.

"coined as a word yet." *The Oxford English Dictionary, Second Edition,* 1989, states that the word *rocketry* was first used by G. E. Pendray in the *Bulletin of the American Interplanetary Society* (November 1930).

"pass beyond the realms of fancy." Forest Ray Moulton, *Astronomy* (New York: MacMillan, 1933) 296.

"fomenting ideas about spaceflight." Frank H. Winter, *Prelude to the Space Age* (Washington: Smithsonian Institution Press, 1983) 15.

"drove men to the moon." Ibid., 113.

"rather less than 10,000 years ago." Maynard Keynes, "Newton the Man," in *Royal Society, Newton Tercentenary Celebrations* (Cambridge: Cambridge University Press, 1947) 27–34.

"genius in the magical field." Parsons, letter to Karl Germer, c.1949, OTO.

1: PARADISE

For a detailed study of Californian life from 1850 to 1950, see Kevin Starr's multivolume *Americans and the California Dream* series.

"the hardest country in the world." Tamar Frankiel, *California's Spiritual Frontiers* (Berkeley: University of California Press, 1988) 8.

Information on the Parsons family, Charles Parsons, author interview, 23 January 2004.

"bandits and desperadoes." Kevin Starr, *Inventing the Dream* (New York: Oxford University Press, 1985) 13.

"the city of Demons." Rev. James Woods, "Los Angeles in 1854–55: The Diary of Rev. James Woods," *Southern California Quarterly* (June 1941), 65–66.

"the serpent also." Frankiel, *California's Spiritual Frontiers,* 59.

"fake healers, Chinese doctors." Steward Edward White, "The Rules of the Game," in *Writing Los Angeles: A Literary Anthology,* ed. David L. Ulin (New York: Library of America, 2002) 29.

"hatred of authority." John Whiteside Parsons, *Analysis by a Master of the Temple of the Critical Modes in the Experiences of His Material Vehicle* (unpublished, c. 1948), YORKE.

"growing urban center of Los Angeles." Ann Scheid, *Pasadena: Crown of the Valley* (Northridge, Calif.: Windsor Publications, 1986) 29.

"inherent love of flowers." Kevin Starr, *Inventing the Dream: California Through the Progressive Era* (New York: Oxford University Press, 1985) 100.

"the most beautiful residence street in the world." *Los Angeles Times,* 20 February 1916.

"Valley Hunt Club." *Pasadena Star-News,* 12 March 1927.

"Jack sitting on her ample knee." Marjorie Cameron, interview by Scott Hobbs and Carol Caldwell, c. 1995, THELEMA.

2: MOON CHILD

"literature and scholarship." Parsons, *Analysis.*

"wanted to go to the moon." Paul Mathison interview by Brad Branson, n.d., ADASTRA.

"that most enticing ingredient—plausibility." Many of Verne's predictions succinctly anticipated the reality of space flight. He named Florida as an ideal launch site, suggested the idea of weightlessness in space, the use of rockets to alter orbit, and an eventual splashdown at sea. Admittedly some of Verne's calculations were awry. It was estimated that the g-force exerted on Verne's gentlemen explorers during blast-off from their cannon would squash them flat within their spacecraft.

"that we are being taught." *Amazing Stories* (April 1926).

"science had always been magic made real." Jack Williamson, *Wonder's Child: My Life in Science Fiction* (New York: Bluejay Books, 1984) 48.

"science fiction fatherhood." Frank H. Winter, *Prelude to the Space Age* (Washington, D.C.: Smithsonian Institution Press, 1983) 25. Description of the history of rocketry from the following: Andrew G. Haley, *Rocketry and Space Exploration* (Princeton, New Jersey: D. Van Nostrand Company, Inc., 1958), Willy Ley, *Rockets, Missiles, and Men in Space* (New York: Signet, 1969); William S. Bainbridge, *The Spaceflight Revolution: A Sociological Study* (Florida: Robert E. Krieger, 1983); Frank H. Winter, *Rockets into Space* (Cambridge, Mass.: Harvard University Press, 1990); Ron Miller, *The History of Rockets* (Danbury, Conn.: Franklin Watts, 1999).

"one of the more popular games." Lloyd Shier, e-mail to author, 15 December 2002.

"Rabelais's unsqueamish stomach." Aldous Huxley, *Los Angeles: A Rhapsody from Jesting Pilate* (New York: George H. Doran, 1926) 302.

"crowd of some 40,000." Kevin Starr, *Material Dreams: Southern California Through the 1920s* (New York: Oxford University Press, 1990), 166.

"mummy's boy." Liljan Wunderman, author interview, 24 January 2003; 24 April 2003.

"Unfortunate experiences with other children." Parsons, *Analysis*.

"closest friendship of Jack Parsons' life." Jeanne Forman, interview by Brad Branson, 15 July 1995, JPL.

"saw the help that he needed." Helen Parsons Smith, author interview, 19–24 March 2003.

"scared himself witless." While Parsons' *The Book of Babalon* (unpublished, c.1946) dates this incident to his thirteenth year, in *Analysis by a Master of the Temple* he claims it took place at the age of sixteen.

"kind of adventurous person." Jeanne Ottinger, author interview, 5 April 2004.

"all sorts of explosive stuff." Marjorie Zisch, author interview, 10 December 2002.

"they got a donkey." Helen Parsons Smith, author interview, 19–24 March 2003.

"glowing, piercing eyes." Robert Rypinski, oral history interview by R. Cargill Hall, 11 February 1969, JPL.

"high altitude research." Milton Lehman, *This High Man: The Life of Robert H. Goddard* (New York: Farrar, Straus, 1963) 31–33.

"rockets in an appropriate manner." Ibid., 73–74.

"destructed a house of police." Winter, *Prelude to the Space Age,* 43.

"horsing around with the same stuff." Jeanne Forman, interview by Brad Branson, 15 July 1995, JPL.

"single most glittering event." Starr, *Material Dreams: Southern California Through the 1920s*, 118–19.

"catch himself a showgirl." Helen Parsons Smith, author interview.

"plunged to their death." Kevin Starr, *The Dream Endures: California Enters the 1940s* (New York: Oxford University Press, 1997) 165.

"too complete an identification." Parsons, *Analysis*.

"These young people have an immense energy." *Los Angeles Times*, 8 December 1929.

"the personality of the teacher." Russell Richardson, *Education in 1933*, Pasadena Museum of History.

"the contact with reality." Parsons, *Analysis*.

"a warm gem-like flame." Robert Rypinski, oral history interview.

"the iron must come out." John Whiteside Parsons, letters to Helen Northrup, 31 July 1934–3 September 1934. STARR.

"grandmother to think about." Parsons would eventually be admitted to the University of Southern California in Los Angeles to take two night courses in chemistry, but he would not be a great success. His attendance was sporadic and he gained only a C after his first year, by which time his rocket studies at Caltech had come to the fore.

"what went on in a chemical reaction." Rypinski, oral history interview.

"a 'global' grasp." Frank Malina, oral history interview by James H. Wilson, 8 June 1973, JPL.

"do some calculating for them." Rypinski, oral history interview.

"It was a lousy planet." Winter, *Prelude to the Space Age*, 77.

3: ERUDITION

"a near-religious dedication." Starr, *The Dream Endures*, 62.

"but without pants." William A. Dodge et al., *Legends of Caltech* (California Institute of Technology, 1982) 16.

"must be the smartest students." William Sears, *Stories from a 20th Century Life* (Stanford, Calif.: Parabolic, 1994) 49.

"a technical school of the highest class?" Judith Goodstein, *Millikan's School: A History of the California Institute of Technology* (New York: W.W. Norton & Company, 1991) 47.

"tell you about in an hour." Ibid., 72.

"What man can imagine, he can do." Frank Malina, *The GALCIT Rocket Research Project, 1936–1938*, first presented at the 18th International Astronautical Congress, Belgrade, September 1967; Frank Malina Collection, CALTECH. Also appears in *First Steps Towards Space*, 1985, 113.

"flexible sort of attitude." Malina, oral history interview by Wilson.

"liquid propellant rocket engine." *The GALCIT Rocket Research Group*, 113.

"The great Robert Goddard." Goddard had in fact got a liquid fueled rocket to reach 7,500 feet (1½ miles) in 1935. However, fearful of both derision and the theft of his ideas, he kept this secret until his Smithsonian Institute report *Liquid-Propellant Rocket Development* alluded to it in 1936.

"long shadow." Frank Malina, *Excerpts from Letters Written Home by F. J. Malina on Rocket Research at the California Institute of Technology between 1936 and 1947,* Frank Malina Collection, CALTECH.

"In desperation." Theodore von Kármán, with Lee Edson, *The Wind and Beyond* (Boston: Little, Brown, 1967) 235.

"freewheeling brain." Frank Malina, oral history interview by Wilson.

"Irish blue eyes." Phyllis Seckler, "Jane Wolfe: The Sword: Hollywood," in *In the Continuum,* Volume III, No. 7 (Oroville, California: College of Thelema, 1985) 38.

"I was a teaser if you will." Helen Parsons Smith, author interview.

"even when he tested rockets." Ibid.

"without the vest." Liljan Wunderman, author interview.

"how much I loved you." John Whiteside Parsons, letter to Helen Northrup, 11 August 1934, STARR.

"Oh, where's Helen." Helen Parsons Smith, author interview.

"more scared using it than losing my ring." Helen Parsons Smith, interview by Michael Bloom, 22 November 1998, OTO.

"Perhaps its some trick." John Whiteside Parsons, letters to Helen Northrup, 31 July–3 September 1934. STARR.

"I turned my face to him and nothing happened." Helen Parsons Smith, author interview. Martin Starr provides an alternative reason for the missed kiss in *The Unknown God: W. T. Smith and the Thelemites* (Bolingbrook: The Teitan Press, Inc., 2003). He states that Jack and Helen were not fond of public displays of affection and thus chose not to kiss beforehand.

"He just wanted to work on rockets." Robert Rypinski, oral history interview.

"They were dangerous places." Helen Parsons Smith, author interview.

"happy but haphazard." Starr, *The Unknown God,* 257.

"Even Malina felt rather uncomfortable." Frank Malina, *The GALCIT Rocket Research Project, 1936–1938,* 120.

"reasonableness of the capitalistic system." Malina, *Letters,* 8 March 1936.

"they didn't make any sense to me." Frank Malina, oral history interview by James H. Wilson, 8 June 1973, JPL.

"The three men and Helen also tried marijuana." Helen Parsons Smith, interview with Michael Bloom, 22 November 1998, OTO.

"remainder of the concert in style." Malina, *Letters,* 21 February 1936.

"a pair of good intelligent friends." Ibid., 18 July 1936.

4: THE SUICIDE SQUAD

"nothing was heard of their recent developments." Winter, *Prelude to the Space Age,* 49.

"the only practicable approach to space travel." Ibid., 19–53.

"become something in the nature of a race." Lehman, *This High Man,* 123.

"I am convinced it is a hopeless task." Malina, *Letters,* 29 June 1936.

constantly putting his nose down to the paper." Frank Malina, oral history
erview by Wilson.

"Their attitude is symptomatic." Malina, *The GALCIT Rocket Research Project, 1936–1938,* 114.

"I am rather reluctant to specify." Lehman, *This High Man,* 191.

"What jet velocities." John Whiteside Parsons, letter to Frank Malina, c. 15 August 1936, Theodore von Kármán Collection, CALTECH.

"appeared to suffer keenly." Frank Malina, *The GALCIT Rocket Research Project, 1936–1938,* 117.

"took himself too seriously." Kármán and Edson, *The Wind and Beyond,* 240–242.

"it sounded like fun." Apollo Smith, oral history interview by Dick House, 23 September 1996, JPL.

"most likely 'borrowed' from GALCIT." Clayton R. Koppes, *JPL and the American Space Program: A History of the Jet Propulsion Laboratory* (New Haven: Yale University Press, 1982) 5.

"the test was successful." Malina, *Letters,* 1 November 1936.

"behind their sandbags and cheered wildly." Ibid., 29 November 1936.

"definite delusions and hallucinations." Taken from the Clinical Record of Marvel H. Parsons, St. Elizabeths Hospital, F.S.A, 10 May 1950.

"It was a bit of a shock." John Whiteside Parsons, letter to Helen Parsons, 26 March 1943, STARR.

"cut throats for." Ibid.

"for a few hundred dollars." Helen Parsons Smith, author interview.

"work with me to make it soon." John Whiteside Parsons to Helen Parsons Smith, 9 September–15 September 1937, STARR.

"Son of Heaven." Iris Chang, *Thread of the Silkworm* (New York: Basic Books, 1995) 78–79.

"really arrogant." Apollo Smith, oral history interview.

"he didn't ostracize himself from them." Frank Malina, oral history interview by Wilson.

"five dollar bills wrapped in newspaper." Although $500 is the amount that Malina remembered in his memoir on the GALCIT Rocket Research Project, his letters from the time state that Arnold originally brought them the still sizeable but lesser amount of $100.

"flabbergasted." Malina, *The GALCIT Rocket Research Project, 1936–1938,* 120.

"He told me I was a bloody fool." Frank Malina, oral history interview by Mary Terrall, 14 December 1978, CALTECH.

"great fun ... head scratching." Malina, *Letters,* 26 July 1937; 1 August 1937.

"distance the pendulum swung." Despite its seeming impracticality, the test had been developed by Friedrich Sander in Germany and had also been used by Robert Goddard.

"The Suicide Squad." Malina, *The GALCIT Rocket Research Project, 1936–1938,* 121. While Theodore von Kármán refers to the group of young rocketeers as "The Suicide Club" in his autobiography, *The Wind and Beyond,* I have chosen to use Frank Malina's term "The Suicide Squad."

"a very strange bunch." Hans Liepmann, author interview, 6 December 2002.

"The $1000 fund is rapidly diminishing." Malina, *Letters,* 19 February 1938.

"the department fantasy expert." Ibid. 3 September 1937.

"prohibited slot machines were scattered across the city." Kevin Starr, *The Dream Endures: California Enters the 1940s* (New York: OUP, 1997) 167–8.

"Parsons was their best bet." Helen Parsons Smith, author interview.

"caused turmoil at the trial." *Los Angeles Times*, 10 May 1938.

"contained explosives or not." *The Pasadena Post*, 10 May 1938.

"the young expert." *Pasadena Star-News*, 10 May 1938.

5: FRATERNITY

"FASCISM OR COMMUNISM." *The California Tech*, 27 May 1937.

"The Nazis are really devils!" William Rees Sears, *Stories from a 20th-Century Life* (Stanford: Parabolic Press, 1993) 96.

"he was something of a Renaissance man." Chang, *Thread of the Silkworm*, 80.

"he was known for his pro-Soviet harangues." Sidney Weinbaum, oral history interview by Terrall.

"he deals in chemistry and communism." Hans Liepmann, author interview.

"organize a Communist group at Caltech." Frank Oppenheimer, oral history interview by Judith R. Goodstein, 16 November 1984, CALTECH.

"they felt for the people." Wunderman, author interview.

"townspeople congregated at Weinbaum's house." Frank Malina, Federal Bureau of Investigation file.

"He was soon providing the students with books and pamphlets." John Whiteside Parsons, Federal Bureau of Investigation file.

"political romantic." Frank Malina, oral history interview by Terrall.

"joined the party under a pseudonym, as did Tsien." For a fuller discussion of Tsien's involvement, see Chang, *Thread of the Silkworm*.

"the unifying thing with our group was rockets." Malina, oral history interview by Wilson.

"teenagers 'chiseled' while noone was looking." Walter Daugherty, author interview, 10 March 2003.

"imagination instead of the telescope." Jack Williamson, author interview, 16 September 2002.

"some had even corresponded." Walter Daugherty, author interview.

"Assorted Services." The science fiction author Robert Heinlein would write a short story inspired by Ackerman's company entitled "We Also Walk Dogs," *Astounding Science Fiction* (July 1941).

"my god and my religion." Forrest J. Ackerman, author interview.

"awe-inspiring examples of his skills at hypnotism." Ibid. "He was a super hypnotist," remembered Ackerman; "He hypnotized about everybody in the club except me. I never wanted to lose control of myself, and I remember one member coming round and showing us the little kangaroo that he said he had hopping around in his hand."

"a young Ray Bradbury." Bradbury would go on to write such seminal science fiction books as *The Martian Chronicles* (Garden City, N.Y.: Doubleday, 1950) and *Fahrenheit 451* (Garden City, N.Y.: Doubleday, 1953).

"asked him a barrage of questions." Ackerman, author interview.

"the rocket society the gentleman spoke of." Ray Bradbury, letter to Frank Malina, 31 August 1980, Malina Collection, CALTECH.

"the rocket motor starts functioning." *Pasadena Post,* 15 July 1938.

"looking for the spectacular." Malina, *Letters,* 7 June 1938.

"better imaginations than we do." Ibid., 5 July 1938.

"The Human Rocket." Malina Collection, CALTECH.

"develop better munitions." Malina, *Letters,* 15 May 1938.

"He is beginning to run short of money." Ibid., 1 August 1938.

"completely vanished." Malina, *The GALCIT Rocket Research Project, 1936–1938,* 124.

"Parsons is doing some experimenting." Malina, *Letters,* 19 February 1939.

6: THE MASS

"A gong sounded." Stuart Timmons, *The Trouble with Harry Hay* (Boston: Alyson, 1990) 76. For more background on the characters and internal politics of the Agape Lodge, see Phyllis Seckler, "Jane Wolfe: The Sword: Hollywood," from *In the Continuum* Volume 3, No. 6–10 (Oroville, California: College of Thelema, 1984–86) and *In the Continuum* Volume 4, No. 1–10 (Oroville, California: College of Thelema, 1987–91). Also see Martin Starr's *The Unknown God.*

"We're all going upstairs." Wunderman, author interview.

"taken a shine to Helen." Starr, *The Unknown God,* 257.

"like real water to a thirsty man to Jack." Robert Rypinski, oral history interview.

"social misfits at the time." For a fuller picture of Wilfred T. Smith, see Starr *The Unknown God.*

"Explain me the riddle of this man." Crowley's description is drawn from the following: Charles Richard Cammell, *Aleister Crowley: The Man, The Image, The Poet* (London: Richards Press, 1951); Aleister Crowley, *The Confessions of Aleister Crowley: An Autohagiography,* eds. John Symonds and Kenneth Grant (New York: Hill and Wang, 1969); John Symonds, *The King of the Shadow Realm—Aleister Crowley: His Life and Magic* (London: Duckworth, 1989); Martin Booth, *A Magick Life: A Biography of Aleister Crowley* (London: Hodder and Stoughton, 2000); Lawrence Sutin, *Do What Thou Wilt: A Life of Aleister Crowley* (New York: St. Martin's Press, 2000).

"his personal spiritual mediums." The "Scarlet Woman" referred to the New Testament's Revelation, specifically the woman "arrayed in purple and scarlet." This woman rode on the back of a "great beast." As such, it seemed a suitable term for Crowley's own consort.

"the age of the child." A concise description of Crowley's cosmology has been given by John Symonds in his introduction to Crowley's *The Confessions of Aleister Crowley* (New York: Hill and Wang, 1969): "There have been, as far as we know, two aeons in the history of the world. The first, that of Isis, is the aeon of the woman; hence matriarchy, the worship of the Great Mother and so on. About 500 B.C. this aeon was succeded by the aeon of Osiris, that is the aeon of the man, the father, hence the paternal religions of suffering and death—Judaism, Buddhism, Christianity and Mohammedanism. This aeon came to an end in 1904 when Aleister Crowley received *The Book of the Law*, and the new aeon, that of Horus the child, was born. In this aeon the emphasis is on the true self or will, not on anything external such as gods and priests. The choice of Egyptian names for the aeons is purely arbitrary."

"any attempt on his leadership was doomed to failure." Starr, *The Unknown God*, 113.

"propound his religion of Thelema across the world." The A∴ A∴ stressed individual magic attainment on a personal basis rather than among a group. It existed side by side with the OTO; thus, becoming a member in the OTO often meant becoming a member in the A∴ A∴

"persistent use of drugs." Booth, *A Magick Life*, 435–6.

"frigged her for human kindness' sake." Aleister Crowley, personal diary, 1 August 1940, YORKE.

"bankrupting him still further." Martin Starr, *The Unknown God*, 77.

"no hope, no bloody nothing." Crowley, personal diary, 5 September 1943.

"it will clutch at a straw." Aleister Crowley, letter to Wilfred T. Smith, 6 August 1931, YORKE.

"new age in religion." Sydney Ahlstrom, *A Religious History of the American People* (New Haven: Yale University Press, 1972) 1026.

"perfect voice and perfect speaking." Malinda Carter, "Christ Method of Healing or Thought-Transference—Which?" *Harmony* No. 12 (May 1900) 237.

"good and the true." John W. Parsons, letter to Aleister Crowley, 26 November 1943, STARR.

"*The Waste Land* resounds with its ideas." See John B. Vickery, *The Literary Impact of "The Golden Bough"* (Princeton: Princeton University Press, 1973).

"a mistaken association of ideas." Sir James George Frazer, *The Golden Bough—A Study in Magic and Religion* (New York: Macmillan, 1942) 49–50.

"the swarming maggots of near-occultists." Crowley, *The Confessions of Aleister Crowley*, 771.

"handed to you on a tray." Aleister Crowley, letter to Wilfred Smith, 6 June 1928, OTO.

"We are gradually building up a mailing list." Wilfred Smith, letter to Aleister Crowley, April 1934, OTO.

"captains of industry." Aleister Crowley to Wilfred Smith, 6 August 1931, OTO.

"communists and pacifists, or both." Wilfred Smith to Aleister Crowley, 6 May 1935, OTO.

"attractive male aspirants, joining in at times." As well as his numerous affairs with women later in life, it seems possible that Parsons might also have experimented with homosexuality. Although he would speak later in life of "his repressed homosexual component," little evidence could be found to support this point. Paul Mathison, a friend of Parsons' second wife, would claim to have had a brief relationship with Parsons in the early 1950s. However his testimony has been met with extreme skepticism by others who knew both him and Parsons during the period.

"the contrapuntal themes he was asked to play." Timmons, *The Trouble with Harry Hay,* 76.

"the official announcer of the city's annual Tournament of Roses." Starr, *The Unknown God,* 251.

"Parsons visiting only once more that year." Ibid., 258.

"to correlate with the work of the 'quantum' field folks." Seckler, *In the Continuum,* III.7, 40.

"delayed for a while." Wilfred Smith, letter to Aleister Crowley, 20 June 1939, OTO.

7: BRAVE NEW WORLD

"What seems utterly impossible today may be commonplace tomorrow." Sam Moskowitz, *The Immortal Storm: A History of Science Fiction Fandom* (Westport, Conn.: Hyperion, 1974) 220.

"along democratic, impersonal, and unselfish lines." John Bristol, *Fancyclopedia* (Los Angeles: F. J. Ackerman, 1944) 74.

"a moment of immortal sadness." Ackerman, author interview.

"be paid for doing our rocket research." Frank Malina, *The U.S. Army Air Corps Jet Propulsion Research Project, GALCIT Project No. 1, 1939–46: A Memoir,* in *Essays on the History of Rocketry and Astronautics: Proceedings of the Third Through Sixth History Symposia of the International Academy of Astronautics,* R. Cargill Hall, ed. (Washington: National Aeronautics and Space Administration, 1977).

"afford smoking ready-rolled cigarettes." Malina, *Letters,* 13 March 1939.

"unnerving explosions." Kármán, *The Wind and Beyond,* 242.

"such a thing as rockets." Ibid., 243.

"had a mission to perform." Malina, oral history interview by Wilson.

"they hadn't gone to university." Jeanne Forman, interview by Brad Branson, 15 July 1995, JPL.

"in the worst sense of the word." Malina, *Letters,* 7 April 1940.

"staff members for the night." Ibid., 21 October 1939.

"gas tight union." Frank Malina, John Parsons, Edward Forman, *Air Corps Jet Propulsion Research, GALCIT Project No. 1, Report No. 3, Final Report for 1939–40, June 15, 1940,* JPL. For a fuller discussion of the GALCIT Project's rocketry work see Benjamin S. Zibit, *The Guggenheim Aeronautics Laboratory at Caltech and the Creation of the Modern Rocket Motor (1936–1946): How the Dynamics of Rocket Theory Became Reality* (unpublished dissertation, City University of New York, 1999), JPL.

"The Crucible of Power." *Astounding Science Fiction* (February 1939).

"dizzy from instructions." Malina, *Letters,* 14 August 1939.

"People will get used to it." Ibid., 20 October 1939.

"too early for Parsons and Forman." Ibid., 21 August 1939.

"we will tell Parsons to keep on trying." Kármán, *The Wind and Beyond,* 245.

"burning would remain a stable process." Zibit, *The Guggenheim Aeronautics Laboratory,* 343.

"Tho the road is long and dreary." John Whiteside Parsons, untitled poem, September 1940, STARR.

"No serious damage, just about a week's worth of repair." Malina, *Letters,* 4 February 1940.

"the group was jubilant." Kármán, *The Wind and Beyond,* 246.

"spies under every piece of paper." Malina, *Letters,* 3 September 1940.

"This isn't the type of problem." Ibid., 20 August 1940.

"an awful lot of time together." Malina, oral history interview by Wilson.

"Malina could be found poring over." Ibid.

"I just couldn't take seriously." Ibid.

"I have set my eyes on high." John Whiteside Parsons, letter to Helen Parsons, 26 March 1943, STARR.

"like a costume party." Liljan Wunderman, interview by Benjamin Zibit, 22 April 1996, JPL.

"I have to tell Frank about this." Liljan Wunderman, author interview.

"Darker Than You Think." "Darker Than You Think" appeared in *Unknown* (December 1940).

"at least with skeptics." Jack Williamson, author interview.

"Satan was not invoked." Jack Williamson, *Wonder's Child: My Life in Science Fiction* (New York, N.Y.: Bluejay Books, 1984) 128.

"fancy Russian cigarettes." Grady McMurtry, letter to Parsons, 8 May 1943, OTO.

"talked about rockets, witchcraft." Grady McMurtry, personal diary, OTO.

"JP is going to be very valuable." Wilfred Smith, letter to Crowley, 21 March 1941, OTO.

"attainment as a matter of fact." Jane Wolfe, letter to Karl Germer, 8 April 1941, OTO.

"don't pay attention to him." Liljan Wunderman, author interview.

"it is nice to have one now." John Whiteside Parsons, letter to Wilfred T. Smith, 3 May 1942, STARR.

8: ZENITH

"who came closest without hitting won." Homer J. Stewart, interview by Susan Pile and Brad Branson, n.d, ADASTRA.

"nothing was impossible." Jeanne Ottinger, author interview, 5 April 2004.

"they were crazy, what could you do?" Apollo Smith, oral history interview.

"Le Page's all-purpose stationery glue." Homer A. Boushey, *A Brief History of the First US JATO Flight Tests of August 1941: A Memoir,* in History of Rocketry and Astronautics. IAA Simposia. Vol. 8. AAS History Series. Vol. 14. (San Diego, Calif.: Univelt, 1993).

"anyone who could drive a car could fly it." Ibid. After the war the Ercoupe would become one of the most popular privately owned airplanes. Its lack of control pedals caused the creation of a new type of pilot's license and it could be found on sale in such unlikely places as the menswear section of Macy's department store.

"It couldn't have been any worse." Kármán, *The Wind and Beyond*, 249.

"The phase of research that was supposed to be solved." Malina, *Letters*, 16 July 1941.

"Well at least it isn't a big hole." Kármán, *The Wind and Beyond*, 250.

"A Super-8 film made of the tests." Film courtesy of JPL/ADASTRA.

"None of us had ever seen a plane climb." Kármán, *The Wind and Beyond*, 250. It was not, however, the world's first rocket-assisted flight. Fritz von Opel, the son of the German automobile pioneer, claimed to have made a rocket-boosted glider flight in 1929 in Frankfurt, Germany, and the Russians had also independently achieved rocket-assisted glider flight in 1940.

"an increase of 56 percent." Frank Malina and John Parsons, *Results of Flight Tests of the Ercoupe Airplane with Auxiliary Jet Propulsion Supplied by Solid Propellant Jet Units*, Galcit Project No. 1, Report No. 9, JPL.

"pulled out of the shoulder socket." Boushey, *A Memoir*.

"an airplane without a prop." Fred Miller, *First JATO: Ammonium Nitrate, Constarch, Black Powder and Glue*, JPL.

"striving for the past three years." Malina, *Letters*, 20 August 1941.

"the squad belonged to the Communist Party." Weinbaum, oral history interview.

"night watchman was in jail." Phyllis Seckler, letter to the author, 11 January 2003. In his memoirs, Malina recalls that Paul Seckler worked as a mechanic.

"Malina thought a séance." Frank Malina, oral history interview by Wilson.

"fits of drunken violence." Seckler, *In the Continuum*, III.7, 33.

"I didn't really know all that was going on." Frank Malina, oral history interview by Wilson.

"give the Fascists hell." Malina, *Letters*, 22 March 1942.

"The rocketeers wondered aloud whether." Malina, *A Memoir* (Vol. 2).

"the exploitation of our ideas." Ibid.

"that dread brother of learning, industry." Frank Malina, oral history interview by Wilson.

"go into manufacturing in any case." Kármán, *The Wind and Beyond*, 256.

"the radio business in the late 1930s." Various, *Aerojet: The Creative Company*, (Los Angeles: Stuart F. Cooper Company, 1995).

"produce JATOs for the armed forces." Towards the end of 1941 another rocket project was instigated on campus at Caltech. Under the direction of Charles Lauritsen, thousands of barrage rockets—explosive rockets designed to be fired from land, air, or sea—were tested and produced at Caltech. Many thousands were made and used in both Europe and the Pacific. For more information see Judith Goodstein, *Millikan's School*, 239–260.

"without blowing up anything." Walter B. Powell, oral history interview by James Wilson, 29 March 1973, CALTECH.

"a teacher, and some rocket tinkerers." It has been suggested by some sources that none of the founders, with the exception of the lawyer Haley, actually possessed the money to start the company, leaving Haley to put up the entire $2,500.

"Aerojet did not even make the business pages." Various, *Aerojet: The Creative Company*.

"Chinese stuff." Kármán, *The Wind and Beyond*, 261.

"an ulcer from worry." Various, *Aerojet: The Creative Company*.

"This is the crossroads." Ibid.

"Andrew G. Haley stood for 'God'." Apollo Smith, oral history interview.

"He claims to hold 6 secret patents." Parsons, letter to Frank Malina, 4 May 1942, JPI.

"in the back of Jack's mind." Frank Malina, oral history interview by Wilson.

"a collaborative effort." Parsons was eventually recognized by the patent office as the sole inventor of the fuel, and a patent was granted in his name.

"Greek Fire." As well as listing other more pedestrian uses of asphalt, Parsons mentions Greek Fire in his report written with Mark Mills, *The Development of an Asphalt Base Solid Propellant*, Air Corps Jet Propulsion Research, GALCIT Project No. 1, Report Nos. 1–15, October 16, 1942, JPL.

"Why not get rid of black powder." It has been suggested by the late Dr. John Bluth of the Jet Propulsion Laboratory Archives that Parsons may have been equally inspired by the form in which roofing tar was delivered—in long cylindrical tubes of tar wrapped in paper, not unlike giant black cigars.

"then added potassium perchlorate." Almost as radical as his decision to use asphalt as a fuel was his use of potassium perchlorate as an oxidizer. The army ordnance department had previously placed a ban on all uses of this chemical as being too toxic. Parsons knew that this toxicity had largely been due to the impurities of the chemical. Knowing that improvements in chemical manufacturing had largely rid the chemical of its toxicity, he did not care to wait for the regulations to be changed before experimenting with it.

"much safer to handle." Parsons and Mills, *The Development of an Asphalt Solid Propellant*.

9: DEGREES OF FREEDOM

"picking up pointers." For a fuller exploration of the growing military interest and different rocket ventures started throughout the war, see Andrew Haley's *Rocketry and Space Exploration*.

"several new angles on storage." Parsons, letter to Helen Parsons, 1 May 1942, HPS.

"an exalting experience." Ibid.

"she thinks she's so right." Jane Wolfe, personal diary, 15 September 1942, OTO.

"as if she owned them." Helen Parsons Smith, author interview.

"with my own husband." Starr, *The Unknown God*, 256.

"combats with Jack." Seckler, *In the Continuum*, III.7, 38.

"and would do it again." Parsons, letter to Helen Parsons, 25 June 1941, HPS.

"I prefer Betty sexually." Ibid., 29 March 1943.

"Georgia calls to Roy." Max Schneider, personal diary, 6 July 1943, OTO.

"from pinnacle to pinnacle." John Whiteside Parsons, letter to Helen Parsons, 2 May 1942, STARR.

"confirmation in the law of Thelema." Parsons, *Analysis*.

"the sore spot I carried." Starr, *The Unknown God,* 271.

"Sex before going up to Mass." Agape Lodge minutes, 15 April 1942, OTO.

"visibly moved by his words." Jane Wolfe, letter to Karl Germer, 26 March 1942, OTO.

"needs a Guru." Germer, letter to Wolfe, 12 March 1942, OTO.

"What heavily restricted Orange Grove." Wolfe, letter to Germer, 8 June 1942, OTO.

"much to the landlord's alarm." Helen Parsons Smith, author interview.

"secret passage." Nieson Himmel, interview by Susan Pile and Brad Branson, n.d., ADASTRA. Description of 1003 Orange Grove drawn from Seckler, *In the Continuum,* III.8, 34–38.

"hold 50 people easily." Jane Wolfe, letter to Karl Germer, 28 June 1942, OTO.

"there was a knife in the room." Robert Cornog, interview by Scott Hobbs, n.d., THELEMA.

"the garden was enchanted." Jeanne Ottinger, author interview.

"screws, electric connections, sawing." Seckler, *In the Continuum,* III.8, 34–38.

"into the foulest of moods." Phyllis Seckler, letter to the author, 11 January 2003.

"I'm a great one to talk," John Whiteside Parsons, letter to Edward Forman, 26 April 1944, private collection.

"outsiders in the field." Jane Wolfe, letter to Karl Germer, 10 March 1943, OTO.

"It was a time in society." Jeanne Forman, interview with Brad Branson, 15 July 1995, JPL.

"not cause a ripple in the public consciousness." Jane Wolfe, letter to Karl Germer, 2 December 1942, OTO.

"coked up like a snowbird." Grady McMurtry, letter to Parsons, 8 May 1943, OTO.

"amphetamine abuse." Nieson Himmel, interview.

"Jack was an expert on drugs." Robert Cornog, interview by Brad Branson and Susan Pile, 7 March 1995, ADASTRA.

"I am no prude." Karl Germer, letter to Aleister Crowley, c. 1943, OTO.

"This will I think rank as the most important thing." Aleister Crowley, personal diary, 19 August 1942, YORKE.

"some definite personal action." Jane Wolfe, letter to Wilfred Smith, 13 January 1943, OTO.

"Love and trust. Jack and Helen." John Whiteside Parsons and Helen Parsons, telegram to Aleister Crowley, 27 January 1943, YORKE

"drug-traffic and so on." Aleister Crowley, letter to Jane Wolfe, 4 May 1943, OTO.

"Jack's trouble is his weakness." Aleister Crowley, letter to Jane Wolfe, December 1943, OTO.

"to ensure his allegiance." Karl Germer, letter to Jane Wolfe, 26 September 1943, OTO.

"the publication of the tarot." Karl Germer, letter to Grady McMurtry, 6 April 1946, OTO.

"he has poise and dignity." Max Schneider, personal diary, 22 June 1943, OTO.

"push on to the Third Degree." Agape Lodge minutes, 1 June 1943, OTO.

"get rid of the old influence." Crowley, letter to Max Schneider, 6 July 1943, YORKE.

"Many times these many years." Wilfred Smith, letter to Crowley, 14 September 1943, YORKE.

"This is no place for indigent mothers." Quoted in Roy Leffingwell, letter to Crowley, 9 April 1944, STARR.

"appalling egotism, bad taste." Parsons, letter to Crowley, 14 September 1943, STARR.

"getting rid of the smoke." Kármán, *The Wind and Beyond*, 264.

"from 10 to 12 hours daily." Malina, *Letters*, 20 November 1942.

"keen to be taken on as consultants." Fritz Zwicky, oral history interview by R. Cargill Hall and James H. Wilson, 27 April 1972, JPL.

"the most frightening experiences of their lives." Gene Pierce, oral history interview by James H. Wilson, 19 June 1972, JPL.

"That, I suppose, represents some progress." Frank Malina, *Letters*, 3 August 1943.

"underneath a nearby tree." Martin Summerfield, oral history interview by J. D. Hunley, 27 September 1994, JPL.

"it had just been a joke." Gene Pierce, oral history interview.

10: A NEW DAWN

"controlled U-235 bombs." *Astounding Science Fiction* (March 1944).

"vindicate science fiction's role." Parsons was to have his own run-in with the Manhattan Project. In May 1945 the security division of the Manhattan Project was alerted to the fact that a chemical, or explosive being used in the highly secretive project, known only as "x-metal," had been procured by Parsons and Forman's Ad Astra Research Company. Supposing espionage, the FBI was alerted and Parsons soon found himself being grilled about his usage of the substance. The exact details of Parsons' deposition are unknown; however, the FBI investigation eventually noted that both he and Forman were no threat to national security.

"drink cheap sherry and talk over new stories." Williamson, *Wonder's Child*, 129.

"unsurmised wonders of chemistry and astronomy." Anthony Boucher, *Rocket to the Morgue* (New York: Duell, Sloan and Pearce, 1942) 53; 112–13.

"super-weapons and atom-powered space ships." L. Sprague de Camp, *Time and Chance* (New Hampshire: Donald M. Grant, 1996) 184–7.

"I was aware or more and more." Frank Malina, *America's First Long-Range-Missile and Space Exploration Program: The ORDCIT Project of the Jet Propulsion Laboratory, 1943–46: A Memoir*, in *Essays on the History of Rocketry and Astronautics: Proceedings of the Third Through Sixth History Symposia of the International Academy of Astronautics*, ed. R. Cargill Hall, (Washington, D.C.: National Aeronautics and Space Administration, 1977), 339.

"space work with planes." Wolfe, letter to Crowley, 26 November 1943.

"solely for current and future wars." Malina, *America's First Long-Range-Missile and Space Exploration Program,* 340.

"[it] was the child of war." Zibit, *The Guggenheim Aeronautics Laboratory,* 410.

"He wouldn't settle down." Andrew Haley Jr., author interview, 26 November 2002.

"as if it was a coat." Ibid.

"He had a slight satanic look." Delphine Haley, author interview, 2 December 2002.

"a childish sense of humor." George Frey, interview by Scott Hobbs, c. 1995, THELEMA.

"Get over here right now." Jeanne Forman, interview by Brad Branson, 15 July 1995, JPL.

"*It's like a fairy story.*" Various, *Aerojet: The Creative Company* (San Dimas: A&M Services, 1995).

"a crack-pot." John Whiteside Parsons, Federal Bureau of Investigation file.

"there were several hundred." Malina, *America's First Long-Range-Missile and Space Exploration Program,* 340.

"hadn't been part of the inner circle." Dorothy Lewis, oral history interview by James Wilson, 15 June 1972, JPL.

"enamored with explosives." Fritz Zwicky, oral history interview by Hall and Wilson.

"very excited." Malina, *America's First Long-Range-Missile and Space Exploration Program,* 351.

"It would have been a disaster." Frank Malina, oral history interview by Wilson.

"like Billy Graham." Zwicky.

"very strange odor." Charles Bartley, oral history interview by John Bluth, 3 October 1994, JPL.

"Just a note from the midst." Parsons, letter to Forman, 26 April 1944.

"whether it was wanted or not." Apollo Smith, oral history interview by Jennifer K. Stine, 25 September 1995, CALTECH.

"bring it to the test stage." Minutes of Solid Conference Meetings (GALCIT), 1944, JPL.

"reputable institution." Zwicky.

"probably earned a similar sum." Charles Bartley and Robert Rypinski both state that Forman and Parsons were each paid $50,000 for their stock. Jeanne Forman recalled Ed Forman gaining some $30,000 from the sale of his stocks. Both of these figures seem slightly inflated. Malina sold half his stock to General Tire in December 1944 for around $7,000.

"these playthings down here in the Arroyo." Zwicky, oral history interview by Hall and Wilson.

"listening to his ideas." Rypinski, oral history interview.

"The jig was pretty much up." Bernard Smith, *Some Vignettes from an Early Rocketeer's Diary: A Memoir,* AAS History Series Vol. 12, JPL.

"He liked to spend hours." Robert Cornog, interview by Brad Branson and Susan Pile, n.d., ADASTRA.

"find himself a new job." Seckler, *In the Continuum,* IV.4, 43.

"boom industry, Laundromats." Bartley, oral history interview.

"the party scattered." Cornog, interview by Branson and Pile.

"coffin of the old OTO." Agape Lodge minutes, 6 January 1945, OTO.

"Buddha-like figure." For a fuller biography of Himmel see *Los Angeles Times,* 14 March 1999.

"wilderness is paradise now." Lou Goldstone to Grady McMurty, 10 March 1945, OTO.

"rugged individualist." Robert Cornog, Federal Bureau of Investigation file.

"created quite a flap in Pasadena." Alva Rogers, *Darkhouse,* featured in *Lighthouse* fanzine, 5 February 1962, (courtesy of the Fanzine Collection of the Paskow Science Fiction Collection at Temple University).

"how they were atheists." Himmel, interview by Branson and Pile.

"The professional fortune teller." Rogers, *Darkhouse.*

"compose some new music for it." Parsons, letter to McMurtry, 8 August 1945, OTO.

"they're having a party." Ottinger, author interview.

"Alva never seemed to have money for rent." Himmel, interview by Branson and Pile.

"In the living room." Rypinski, oral history interview.

"I could never have." Himmel, interview by Russell Miller, 14 August 1986. See Russell Miller, *Bare-Faced Messiah: The True Story of L. Ron Hubbard* (London: Michael Joseph, 1987) 112–31. By permission of PFD.

"she 'wasn't sophisticated'." Himmel, interview by Branson and Pile.

"I can still describe Betty's swimsuit." Cornog, interview by Branson and Pile.

"pretty much with an iron hand." Alice Greenberg, author interview, 9 March 2003.

"lack of subtlety and humor." Crowley, letter to Germer, 15 November 1943, OTO.

"certain people to come to the telephone." Seckler, *In the Continuum,* IV.5, 42.

"an ordeal set by the gods." Germer, letter to Wolfe, 12 January 1944, OTO.

"my teacher, my guide." Parsons to Smith, 6 February 1945, STARR.

"Smith is a menace." Seckler, *In the Continuum,* IV.4, 35.

"Smith who has a master hand." Ibid.

"Have I got to explain to everybody." Ibid., 37.

"I will give you a war-engine." Crowley, letter to Louis Wilkinson, 7 August 1945, YORKE.

"became the neighborhood experts." Alva Rogers, *A Requiem for Astounding* (Chicago: Advent, 1964) 138.

"desperate social problems." Malina, *America's First Long-Range-Missile and Space Exploration Program,* 365.

11: ROCK BOTTOM

"almost religious ecstasy." Rogers, *Darkhouse.*

"the model for the charismatic seducer." Williamson, *Wonder's Child,* 129.

"counterintelligence during the war." *L.Ron Hubbard: A Chronicle,* http://
www.scientology.org/html/en_US/l-ron-hubbard/chronicle/pg006.html.

"bullet wounds." Miller, *Bare-Faced Messiah,* 100.

"broken feet," Robert Heinlein, *Agape and Eros: The Art of Theodore
Sturgeon,* preface to Theodore Sturgeon's *Godbody* (New York: Donald I. Fine,
1986) 12.

"he had been sunk four times." Ibid.

"patrolling the frozen Aleutians." Himmel, interview by Miller.

"aboriginal arrows." Rogers, *Darkhouse.*

"measured his skull and declared it to be unique." Himmel, interview by Miller.

"I recall his eyes." Williamson, *Wonder's Child,* 185.

"charm the shit out of anybody." Himmel, interview by Miller.

"I thought he was a bastard." Greenberg, author interview.

"one hell of a good story." Rogers, *Darkhouse.*

"in complete accord with our own principles." John Symonds, *The King of the
Shadow Realm* (London: Duckworth, 1989), 562–3.

"batting ideas back and forth." Greenberg, author interview.

"He was irresistible to women." Himmel, interview by Branson and Pile.

"rapping her smartly across the nose." McMurtry, *Report on the Order in
Southern California,* 25 January 1946, OTO.

"like a starfish on a clam." Cornog, interview by Susan Pile and Brad Branson.

"the genial elder brother." Seckler, *In the Continuum,* IV.5, 42.

"hitherto unfelt passion, jealousy." Rogers, *Darkhouse.*

"right in front of Parsons." Himmel, interview by Miller.

"There is not even any point to it." Crowley, letter to Parsons, 19 October
1943, YORKE.

"I prefer the appearance of evil." Parsons, letter to Crowley, 26 November
1943, STARR.

"Let's work on the heavier stuff." Lynn Maginnis, author interview.

"so long as he got a result." Seckler, *In the Continuum,* IV.4, 41–42.

"alien and inimical." Ibid.

"the adventurer is an individualist." William Bolitho, *Twelve Against the Gods:
The Story of Adventure* (New York: Simon and Schuster, 1929) 1.

"weird and disturbing noises." Rogers, *Darkhouse.*

"I performed certain operations." John Whiteside Parsons, *Introduction to the
Record of the Invocation of Babalon* (unpublished, n.d.), OTO.

"that's not what I asked for." Parsons, letter to Crowley, 21 January 1946,
YORKE.

"struck strongly on the right shoulder." John Whiteside Parsons, *The Book of
Babalon* (unpublished, c. 1950), YORKE.

"Hubbard's right arm remained paralyzed." Ibid.

"Hubbard's supreme magical sensitivity." Ibid.

"It is done." Ibid.

"giant red lips." Maginnis, interview.

"'mad scientist' owner." It has been suggested by Martin Starr that Marjorie

Cameron was dating Alva Rogers, one of the LASFS contingent living in 1003, before she met Parsons.

"I just couldn't wait to get there." Marjorie Cameron, interview by Bill Breeze, n.d., OTO.

"I probably derided it." Ibid.

"two weeks, the couple barely left Parson's room." Ibid.

"beyond most of my actual knowledge." Seckler, *In the Continuum*, IV.4, 43.

"some form of manic episode." see the section on mood disorder criteria in *Diagnostic and Statistical Manual of Mental Disorders: DSM IV* (Washington, D.C.: American Psychiatric Association, 1994).

"performing magic with Parsons for nearly two months." Parsons, *Book of Babalon*.

"pale and sweating." Ibid.

"a fatigued Hubbard." Ibid.

"I must put the Lodge in [other] hands." Parsons, letter to Crowley, 6 March 1946, YORKE.

"In the coming months the world approaches." Parsons, letter to all Agape Lodge Members, 16 March 1946, OTO.

"move out by June 1." Maria Prescott, letter to Grady McMurty, 13 April 1946, OTO.

"To pool and accumulate earnings." *Parsons v. Hubbard and Northrup*, Case No. 101634, Circuit Court, Dade County, Florida, 11 July 1946.

"cement the ménage." Seckler, *In the Continuum*, IV.4, 42.

"Hubbard came up with a proposal." Seckler, *In the Continuum*, IV.5, 49.

"Parsons was easily persuaded." Louis Culling, letter to Karl Germer, 12 May 1946, OTO.

"more of an adventure than a business proposition." Grady McMurtry, letter to Maria Prescott, 27 April 1946, OTO.

"if Ron is another Smith?" Jane Wolfe, letter to Karl Germer, 16 May 1946, OTO.

"to visit Central & South America & China." L. Ron Hubbard, Naval Personnel Record, file number 113392, www.lermanet//-Ron_Hubbard.

"he had no money for supplies." Maria Prescott, letter to Grady McMurty, 13 April 1946, OTO.

"He told friends he was going to persuade." Louis Culling, letter to Karl Germer, 12 May 1946, OTO.

"eating out of Ron's hand." Louis Culling, letter to Karl Germer, 12 May 1946, OTO.

"I hope we shall always be partners." Ibid.

"obvious victim prowling swindlers." Crowley, telegram to Germer, 22 May 1946, YORKE.

"three sailing yachts." *Parsons v. Hubbard and Northrup*, Case No. 101634, Circuit Court, Dade County, Florida, 11 July 1946.

"Hubbard's latest navy disability check." *Claimant's Appeal to Administrator of Veterans' Affairs*, Claim No. 7017422, 4 July 1946. Hubbard gave his return address as P.O. Box 941, Miami Beach, Florida.

"forced him back to port." Parsons, letter to Crowley, 5 July 1946, YORKE.

"waiting for the errant pair." Ibid.

"The last of the OTO lodgers." Seckler, *In the Continuum,* IV.5, 43.

"It seemed like a tombstone." Robert Cornog, Federal Bureau of Investigation file.

"The more complete story of Hubbard." L. Sprague de Camp, letter to Isaac Asimov, 27 August 1946, DE CAMP.

"continues his process of evolution." *Dianetics—The Evolution of a Science* in *Astounding Science Fiction* (May 1950), 43–87.

"Even schizophrenia and criminal behavior could be cured." Ibid.

"grow a new finger." Forrest Ackerman, author interview.

"for only $600." Williamson, *Wonder's Child,* 184.

"the fastest growing movement in the US." *Daily News,* 6 September 1950.

"are central to its success." Critics claim that the Church of Scientology membership numbers are much lower. See Russ Coffey, "It worked for Travolta. Would it work for me?," *The Times* (London), 18 February 2004.

"'black magic' cult." The statement issued by the Church of Scientology in *The Sunday Times,* December 1969, reads: "Hubbard broke up black magic in America: Dr. [sic] Jack Parsons of Pasadena, California was America's Number One solid fuel rocket expert. He was involved with the infamous English black magician Aleister Crowley who called himself 'The Beast 666.' Crowley ran an organization called the Order of Templars Orientalis [sic] over the world which had savage and bestial rites. Dr. Parsons was head of the American branch located at 1003 Orange Grove Avenue, Pasadena, California. This was a huge house which had paying guests who were the USA nuclear physicists working at Cal. Tech. Certain agencies objected to nuclear physicists being housed under the same roof.

"L. Ron Hubbard was still an officer of the US Navy because he was well known as a writer and a philosopher and had friends amongst the physicists, he was sent in to handle the situation. He went to live at the house and investigated the black magic rites and the general situation and found them very bad.

"Parsons wrote to Crowley in England about Hubbard. Crowley 'The Beast 666' detected an enemy and warned Parsons. This is all proven by the correspondence unearthed by The Sunday Times. Hubbard's mission was successful far beyond anyone's expectations. The house was torn down. Hubbard rescued a girl they were using. The black magic group was dispersed and destroyed and has never recovered. The physicists included many of the 64 top US scientists who were later declared insecure and dismissed from government service with so much publicity."

"in the skies over Yugoslavia." *Los Angeles Times,* 12 September 1946; 15 September 1946; 16 September 1946.

"Parsons worked in the laboratory." John Whiteside Parsons, Federal Bureau of Investigation file.

"with his dagger to stop the wind." Marjorie Cameron, interview by Breeze.

"Jack wasn't acting the way he used to." Jeanne Forman, interview by Branson.

"a bit British." George Frey, interview with Scott Hobbs, c. 1995, THELEMA.

"They were two very independent people." Robert Cornog, interview by Scott Hobbs, n.d., THELEMA.

"She was a nice person." Greenberg, author interview.

"Now I am a whip." John Whiteside Parsons, *Desire,* unpublished poem, private collection.

"He began selling." Wolfe, letter to Germer, 12 August 1946, OTO.

"atomic-like explosion." *Los Angeles Times,* 21 February 1947.

"never finished high school." *Los Angeles Times,* 13 March 1947.

"dreams of interplanetary travel were still keen." *Pacific Rockets,* in *Journal of The Pacific Rocket Society* Vol. 2, No. 1 (Summer 1947).

"ostensibly to study art." Wolfe, letter to McMurtry, 8 August 1947, OTO.

"and a phonograph needle." *Los Angeles Mirror,* 20 April 1953.

"My aim is to rebuild myself." Parsons, letter to Crowley, c. 1947, YORKE.

"the usual dose being one-eighth." Booth, *A Magick Life,* 479.

"without a shudder." *Time* magazine, 15 December 1947, 36.

12: INTO THE ABYSS

"science fiction fans with Communist sympathies." Ackerman, author interview.

"he and his wife Liljan were living apart." Frank Malina, letters to Liljan Malina, 18 March 1946 and 21 March 1946, MALINA.

"mysteriously reappear a few days later." Liljan Wunderman, author interview.

"an alleged Communist Party Member." John Whiteside Parsons, Federal Bureau of Investigation file.

"make the change as soon as possible." Parsons, letter to Kármán, 20 September 1948, Theodore von Kármán Collection, CALTECH.

"sitting up on the roof just flying kites." Jeanne Ottinger, author interview.

"He embarked on a series of magick rituals." Parsons, letter to Germer, 19 June 1949, OTO.

"Thoughts of death and suicide possessed him." John Whiteside Parsons, untitled chronology of the "Black Pilgrimage," 17 November 1948, OTO.

"the coming of BABALON be made open." Parsons, *The Book of Antichrist* (unpublished, c. 1949), OTO.

"who agreed to appeal his suspension." George Frey, interview by Scott Hobbs, c. 2003.

"without sufficient cause." Parsons, Federal Bureau of Investigation file.

"several reports." Parsons, letter to Kármán, 15 June 1949, Kármán Collection, CALTECH.

"a number-one rocket engineer." Robert Heinlein, letter to L. Sprague de Camp, 14 June 1949, DE CAMP.

"with a loose door wired shut." De Camp, *Time and Chance,* 212.

"'offering' him Betty back." De Camp, letter to Asimov, 5 August 1949.

"An authentic mad genius if I ever met one." De Camp to Robert S. Richardson, 14 August 1949, DE CAMP.

"a local nobleman and a bullfighter." Paul Mathison, interview by Scott Hobbs, n.d., THELEMA.

"the 'Concrete Castle'." Robert Cornog, interview by Scott Hobbs, n.d, THELEMA.

"she gives the place a nice feminine touch." George Frey, interview by Scott Hobbs, c. 1995, THELEMA.

"It would be the last time he saw his old friend." Malina, oral history interview by Wilson.

"the Committee had information indicating that Frank Malina." *Los Angeles Times,* 15 June 1949.

"Science, that was going to save the world." John Whiteside Parsons, *Freedom Is a Two-Edged Sword and Other Essays* (Tempe, Arizona: New Falcon Publications, 1989) 10.

"Rosenfeld had fallen ill." Parsons, letter to Kármán, 15 June 1949, Kármán Collection, CALTECH.

"dealt with many 'isms' other than Communism." Parsons, Federal Bureau of Investigation file.

"he could no longer hold his prestigious position." Iris Chang, *Thread of the Silkworm,* 150.

"she alerted the security authorities." Parsons, Federal Bureau of Investigation file.

"an odd and unusual pair." Ibid.

"the 'Black Brotherhood'." Frey, interview by Scott Hobbs, c. 1995.

"the jazz legend, Charlie Parker." Frey, interview by Scott Hobbs, c. 2003.

"the police cars kept coming." Greg Ganci and Martin Foshaug, interview by Brad Branson and Susan Pile, 2 September 1995, JPL.

"he was never really very bohemian." Greg Ganci, author interview, 4 August 2003.

"an austere simplicity of approach." Parsons, letter to Cameron, 22 January 1950, YORKE.

"comparative respectability of the explosives business." Parsons, letter to de Camp, 8 May 1951, DE CAMP.

"due to lack of sufficient evidence." Parsons, Federal Bureau of Investigation file.

"hysteria and depressing melancholy." Parsons, letter to Germer, 11 February 1952, YORKE.

"re-establish the ancient glories." Mathison, interview by Brad Branson, n.d., ADASTRA.

"But I shall come back." Jane Wolfe, letter to McMurtry, 10 October 1952, OTO.

"Mr. Parsons is very inventive." Kármán to Ward Jewell, 4 April 1952, Kármán Collection, CALTECH.

"Mexico where life was one big fiesta." Greg Ganci and Martin Foshaug, interview by Brad Branson and Susan Pile, 2 September 1995, JPL.

"one of the most powerful explosives." *Pasadena Independent,* 19 June 1952.

"to start a 'fireworks company' in Mexico." Bartley, oral history interview.

"who would know about it in the long run." Frey, interview by Hobbs, c. 1995.

"accompanied by their twenty-one year old friend." Ganci, author interview.

"To Frey he seemed thrilled." George Frey, author interview, 24 August 2003.

"was in a bit of a hurry to finish." *Los Angeles Mirror,* 18 June 1952.

"he had test tubes." Ganci, author interview, 4 August 2003.

EPILOGUE

"His strong moral self." Roger Malina, author interview.

"Chairman Mao himself." Chang, *Thread of the Silkworm,* 246.

"a totally different person." Lynn Maginnis, author interview.

"physically abusive." Jeanne Ottinger, author interview.

"The screaming of the banshees." Lynn Maginnis, author interview.

"time-traveling big game hunters." L. Sprague de Camp, *A Gun for Dinosaur,* in *Galaxy Science Fiction* (March 1956) 7–25.

INDEX

Made in United States
Troutdale, OR
10/01/2023

13335296R00228